Rare and Exotic Orchids

Rare and Exotic Orchids

Joel L. Schiff

Rare and Exotic Orchids

Their Nature and Cultural Significance

 Springer

Joel L. Schiff
Mairangi Bay, Auckland, New Zealand

ISBN 978-3-319-70033-5   (hardcover)   ISBN 978-3-319-70034-2   (eBook)
ISBN 978-3-319-88879-8   (softcover)
https://doi.org/10.1007/978-3-319-70034-2

Library of Congress Control Number: 2017959931

Cover image: Bulbophyllum virescens (binnendijkii), courtesy Steve Beckendorf.

Printed on acid-free paper

This Springer imprint is published by Springer Nature
The registered company is Springer International Publishing AG
The registered company address is: Gewerbestrasse 11, 6330 Cham, Switzerland

Preface

In 1861, the year that Abraham Lincoln was inaugurated President of the United States, a Tiger orchid was planted across the world at the Singapore Botanic Gardens. It has seen the presidents of many countries come and go over the past 150 years, as that same original orchid is still thriving and currently on display for visitors of the twenty-first century and beyond. Similarly, some years later, *a Bulbophyllum ornatissimum*, purchased at an auction in 1887, found an agreeable home at the Glasnevin National Botanical Gardens in Dublin and has been growing there ever since.

An orchid produces new growth when the old growth withers and dies. Indeed, if properly cared for, this plant can grow indefinitely. This requires that the orchid grower replicate to a certain extent the growing conditions of its native habitat. Sometimes, those conditions are of steamy South American jungles or the montane rainforests of India, the highlands of New Guinea, the tropical dry forests of Mexico, or coastal lowlands of Brazil.

In a sense, orchids represent something we have lost. Many generations of our ancestors once lived in forested regions in close quarters with nature, but even for many of our grandparents, that might be only a distant memory. Yet traces of our natural heritage are still in our DNA, and so providing *Homo sapiens* with green sanctuaries is essential for coping with modern living.

Living as we do in our own urban jungle, many of us will never step foot in the places where such exotic orchids are endemic. Yet through the study and cultivation of such bizarre and beautiful plants, we can still honor the nature of these mysterious locations without ever leaving the familiarity of our own homes and gardens.

Orchids are like no other family of flowering plants. The endless diversity of form, color, and scent of their flowers; the unusual appearance of their vegetative growth – often with pseudobulbs beneath their leaves; and, of course, the unusual habit of a great many orchids (about 70%) found growing on the branches of trees as epiphytes, distinguish orchids as a fascinating group of flowering plants unlike any other on Earth.

For the above reasons, many people speak of growing orchids as they would of an intense addiction. Once you have started with a single plant, you quickly find yourself growing a second one, then three, then four, then before you know it, they have taken over your house and expanded into a separate greenhouse. Or perhaps you are among those who find themselves searching out or posting enumerable photographs of orchids on Facebook, Pinterest, Flickr, or other image and information-sharing websites such as the Orchid Board. In this day and age, you do not even have to grow orchids to become obsessed with them: an image of their exotic beauty alone is enough for some.

This book is intended for those who wish to embark on this long and slippery slope, or who perhaps already know a little about these unusual plants but now wish to deepen their knowledge in matters both technical and cultural. This is not a book about how to grow orchids – there are already many fine books on that subject. It is a book about the compelling and

Fig. 1 *Hummingbird Perched on the Orchid Plant*, painted by American artist Martin Johnson Heade (1819–1904), captures the exotic romance of the jungle (WikiCommons)

Fig. 2 Central Park in New York as viewed from above. The urge to be near nature is always there (Image courtesy Martin St-Amant/Wiki Commons)

complex history, biology, and etymology of these exotic breeds, touching upon a slew of other intriguing subjects along the way. And of course, it is also a book about sex, since orchids have mastered more procreative techniques than are found in the *Kama Sutra*.

This book does not attempt to cover the many thousands of orchid genera and species. Instead, it seeks to paint a portrait of this exotic world through a selection of some of the most interesting and unusual orchids on Earth. Perhaps the reader's favorite exotic orchid is not included. Yet many others will be, so enjoy this fascinating and unfamiliar journey.

Auckland, New Zealand   Joel L. Schiff
2017

Acknowledgments

Many people provided all manner of assistance, and without their support, this book could have never been written. These individuals came from all corners of the globe and were extremely generous in providing images and sharing their orchid knowledge. All the stunning images have been accredited in the text to those who provided them, and this book is therefore a tribute to their work. In addition, I wish to personally thank those who in various ways assisted in improving the text and in answering questions of a technical nature. The orchid world is populated not only with truly remarkable flowering plants, but also with a truly gracious and generous group of human beings.

Technical assistance was rendered in one form or another by:

Dr. Mario Blanco, Lankester Botanical Gardens, University of Costa Rica
Dr. Devangi Chachad, Jai Hind College, Mumbai, India
Carlos Cruz, Glendale California, USA
Dr. Guido Deburghgraeve, Liedekerke, Belgium
Thomas Ederer, Neusiedl am See, Austria
Dr. Anne Gaskett, University of Auckland, Auckland, New Zealand
Professor Steven D. Johnson, University of Kwazulu-Natal, South Africa
Chien C. Lee, Sarawak, Malaysia
Amy Martin, University of Auckland, Auckland, New Zealand
Katy Metcalf, Auckland, New Zealand, who did all the fine artwork
Ron Parsons, South San Francisco, USA
Professor Hannes Paulus, University of Vienna, Vienna, Austria
Dr. Aaron Schiff, Auckland, New Zealand
André Schuiteman, Kew Gardens, London, England
Dr. Kenji Suetsugu, Kobe University, Kobe, Japan
Dr. Gowhar Ahmed Shapoo, University of Kashmir, Srinagar, India
Ross and Susan Tucker, Tuckers Orchid Nursery, Auckland, New Zealand
Professor Zhong-Jian Liu, The National Orchid Conservation Center, Shenzhen, China
Professor Zong-Xin Ren, Chinese Academy of Sciences, Kunming, Yunnan, China
Kevin Western, Western Orchids Laboratories, Blackwood, Australia

Contents

Nothing in science can account for the way people feel about orchids.

Susan Orlean in *The Orchid Thief*

Orchids are the most highly evolved and diverse flowering plant family on Earth. The origins of these plants date back millions of years. Exactly how old the family is, however, was a longstanding subject for debate, given that there were no remains in the ancient fossil record to work with. That is until recently, when scientists conducting research through Harvard University made an extraordinary discovery [1]. In 2005, a fossil of an extinct bee, *Proplebeia dominicana,* was recovered in the Dominican Republic. The bee specimen was dated to be 15–20 million years old (Fig. 1.1).

What is of interest here is not so much the bee but the orchid pollinia attached to the bee's back. This was determined to come from an orchid of the subtribe Goodyerinae and given the botanical name *Mellorchis caribea.* Using some sophisticated methods of analysis based on both the pollinia's morphology and the molecular genetics of related fossil plants, a date for the common ancestor of all present day orchids was calculated to be 76–84 (average 80) million years ago. Without giving an age, Charles Darwin said as much: "all Orchids owe what they have in common to descent from some monocotyledonous plant…" [2].

The above is remarkably consistent with very recent research (2016), by an international collaboration of fifteen "specialists in orchid systematics, phylogenetics, ecology, and biogeography," which came to the rather amazing conclusion that: "Orchids appear to have arisen in Australia between 102 and 120 million years ago … then spread to the Neotropics via Antarctica between 79.7 and 99.5 million years ago, when all three continents were in close contact." [3]

Interestingly, this predates the extinction event of the dinosaurs 66 million years ago. Thus, some ancient orchids must have survived that mass extinction, after which they underwent an explosion of new orchid genera and species.

When we later discuss the crafty means by which some orchids are pollinated, the reader might be tempted to believe that these plants possess some kind of inherent intelligence. Indeed, in Chap. 4 we will see that when it comes to sex, various insects that do have a brain – albeit a small one – are no match for an ingenious orchid and millions of years of evolution. For example, some male insect species are not only duped into "mating" with an orchid and thereby pollinating it but even prefer to mate with the orchid flower instead of with females of their own species. Now that is cunning!

It should be mentioned that while orchids and other plants do not have a brain per se, they do have a very sophisticated *form of awareness* of their environment and in their ability to respond to it [4]. "[P]lants behave like little statisticians, making implicit inferences about their world through changes in their internal states," researchers Paco Calvo and Karl Friston maintain [5]. Whether this amounts to real intelligence or not in part comes down to how liberally one interprets the word "intelligence." However, some biologists do make claims for a *form of plant intelligence* [6] and have even gone on to create the new field of "plant neurobiology." Of course, this must be in a metaphorical sense, as plants do not have neurons, and admittedly, there is considerable controversy attending this whole issue. We will take an anthropomorphic view of orchid biology from time to

© Springer International Publishing AG 2018
J.L. Schiff, *Rare and Exotic Orchids*, https://doi.org/10.1007/978-3-319-70034-2_1

Fig. 1.1 The extinct fossil bee *Proplebeia dominicana* encased in amber with orchid pollinia attached to its back from the pollination of an ancient orchid species (Image courtesy Santiago R. Ramírez, Harvard University/UC Davis)

time, but only in a metaphorical sense in order to stimulate the reader's sense of wonder.

Ancient History

Seemingly the oldest known reference to orchids was made by the deified being and Father of Chinese medicine, Shen Nung (~2695 BC). In his *Materia Medica*, Shen mentions the medicinal properties of the *Dendrobium* orchid species.

Another ancient Chinese reference to orchids themselves (known as "lan") comes from the *I Ching* (*Classic of Changes*), which dates back to between the ten and fourth centuries BC. Indeed, the Chinese philosopher Confucius (551–479 BC) had a special affection for the orchid and made several references to them in his writings:

> *The orchids grow in the woods and they let out their fragrance even if there is no one around to appreciate it. Likewise, men of noble character will not let poverty deter their will to be guided by high principles and morals.*
>
> *If you are in the company of good people, it is like entering a room full of orchids.*
> (Translation by Alice Poon).

Orchids have long been used in traditional medicine in the Himalayan Kashmir region (see Appendix I), and they continue to be used in tradi-

tional medicine in various parts of the world. A very comprehensive treatment of the medicinal use of orchids throughout Asia, India (where orchid extracts have been used for centuries in Ayurvedic medicine), and the Middle East, can be found in the authoritative text *Medicinal Orchids of Asia* by medical doctor Eng Soon Teoh. Only a few highlights are touched upon in the present text.

The European orchid experience originates from the Mediterranean region, where various species of an orchid genus that has ovoid shaped tubers growing beneath the ground are found. We owe its flavorful history to the great Greek scholar, *Theophrastos* (ca. 371–287 B.C.), who in volume nine of his magnificent ten volume *Historia Plantarum* (*Enquiry into Plants*) first connected orchid tubers with sexual enhancement in a way that has bewitched the minds of male western civilization ever since. Indeed, the genus bears the name *Orchis*, the Greek word for testicle. Incidently, these tubers are used to store starch for food during the arid summer months and recent research indicates that there may be some connection with male sexuality as discussed in the sequel regarding recent research into the drink 'salep' (Fig. 1.2).

Of course, one had to be careful which round tuber was utilized, as the larger of the two gave men sexual vigor, whereas the smaller tuber had the opposite effect. The Romans were not to be left out when it came to aphrodisiacs, as Pliny the Elder in his *Historia Naturalis* wrote of the sexual stimulation to be had from the tubers of the so-called *Satyrion* plant. The name derives from the wantonly sexual mythical figure of Satyr from Greek and Roman mythology, who in the latter had horns like a goat. The *Satyrion's* mythical powers of lust have even passed into literature:

> *In the meantime, the satyrion which I had drunk only a little while before spurred every nerve to lust and I began to gore Quartilla impetuously, and she, burning with the same passion, reciprocated in the game. From: Satyricon by Petronius, Ch.26, 1ˢᵗ century A.D.*

The one who gave the *Orchis* its sexual enhancer seal of approval for the next 1500 years was the Greek physician and herbalist Pedanius Dioscorides (ca. 40–90 A.D.) in his *De Materia Medica*. This authoritative pharmacopeia of herbal medicine

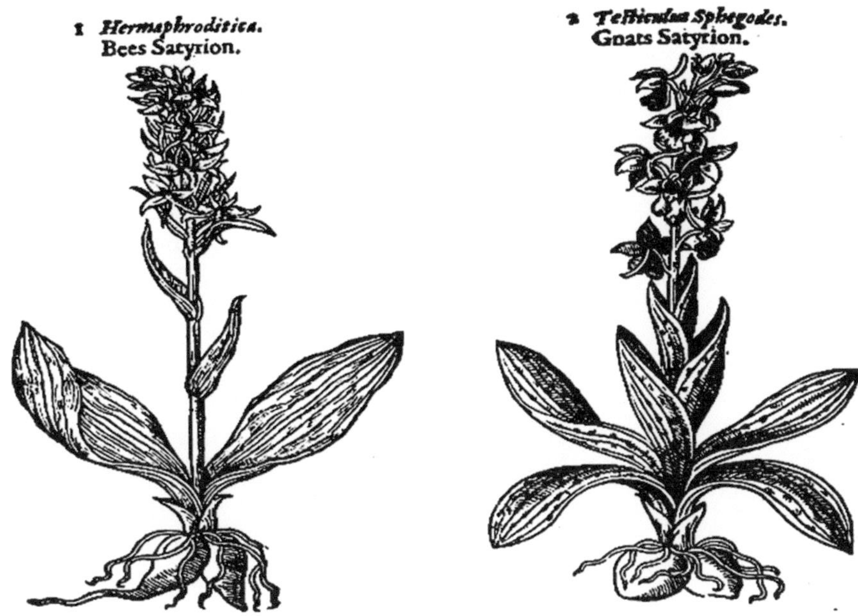

Fig. 1.2 Two species of *Orchis*, known as *Satyrion* in England at the time, from *The Herball, or Generall Historie of Plantes*, by botanist and herbalist John Gerarde (London, 1597)

Fig. 1.3 A depiction of Dioscorides in a thirteenth-century Arabic edition of *De Materia Medica* (Image by permission of The Bodleian Library, University of Oxford, England)

was widely read and consulted during the Middle Ages and beyond (Fig. 1.3).

Not only do the *Orchis* tubers have a male sexual association, but the flowers of the species *Orchis italica* supposedly bear more than a passing resemblance to male figures. According to the *Doctrine of Signatures*, which was widely used among herbalists and supported by religious figures over the centuries, ailments afflicting certain parts of the body were to be treated by herbs that resembled the afflicted part. For example, the flowers of the herb Eyebright (*Euphrasia officinalis*), which look somewhat like eyes, were thought to treat eye complaints such as conjunctivitis, bloodshot eyes, and itching. This notion has even persisted into the twenty-first century regarding herbs such as Eyebright, which, as it so happens, does contain compounds that have been found to reduce inflammation [7]. Could it be that *Orchis italica* itself is expressing the notion that by consuming its tubers one will become similarly endowed? (Figs. 1.4 and 1.5).

Even today, a beverage known as *salep* (or *sahlab*), made from powdered *Orchis* tubers and milk, remains popular throughout the Middle East and India as it has been for centuries. It was even consumed in England in the 18th and 19th centuries, where it was known as Saloop and was less expensive than tea or coffee. According to *A Modern Herbal* by Mrs. Grieve:

> *… it was considered so important an article of diet as to constitute a part of the stores of every ship's company in the days of sailing ships and long voyages, an ounce dissolved in 2 quarts of boiling water, being considered sufficient subsistence for each man per day, should provisions run short [8].*

Fig. 1.4 Perhaps a man will become as equally endowed as the flowers of this *Orchis italica* by consuming its testicular-like tubers (Image courtesy Luis Nunes Alberto/Creative Commons)

Often made from *Orchis mascula* and *Orchis militaris*, salep is a thick coffee substitute still regarded by some as a sexual tonic and prescribed for various sexually related complaints, if no longer for "disorders stemming from acrimony in the juices." Additionally, a salep-flavored ice cream known as *dondurma* is popular in Turkey, and it can even be found in the author's favorite Turkish restaurant down the road.

So strong was the belief that orchids were somehow entwined with sexual matters that by the nineteenth century, the English art critic John Ruskin described their flowers as "prurient apparitions." He would have been referring to their general appearance, which one can interpret as sexually suggestive. It is fair to say however, that Ruskin might have been biased by the times and by a few

sexual issues of his own, having failed to consummate his 6-year marriage. Perhaps a little glass of salep would have helped (Fig. 1.6).

In 2009, an Indian study was conducted in order to investigate the reputed aphrodisiacal properties of salep [9]. In the study, adult male mice (specially bred for research purposes) were split into three groups. The first group was fed a 1% dose of gum acacia in water, the second group received powdered tubers of the medicinal terrestrial orchid *Eulophia nuda* (syn. *Euph. spectabilis*), and a third group was fed the powdered tubers of *Orchis latifolia* (syn. For *Dactylorhiza incarnata*), used for making salep. Note that the related species *Dactylorhiza hatagirea* (syn. *Orchis hatagirea*) appears in Appendix I as an aphrodisiac under Condition #11 (Figs. 1.7 and 1.8).

The results of this scientific study were as follows: No behavioral or physiological change was observed in the control group, but mounting behavior of the male mice was somewhat increased in the *Eulophia nuda* fed group and significantly increased in the *Orchis latifolia* group; increases in testosterone levels in the experimental groups, with up to 20% higher for the mice fed on *Orchis latifolia*; increased sperm count in the experimental groups, again more so with those mice fed *Orchis latifolia*; and lowered cholesterol as an added benefit in the experimental groups. The latter could be due to the substance *glucomannan*, a starchy polysaccharide that is a source of dietary fiber. It has been shown to reduce total cholesterol and may play a role in the treatment of type 2 diabetes. It should be mentioned that while these results seem promising, there are associated health risks with currently available glucomannan supplements, and so these supplements should be avoided. While animal studies are not always predictive of human response, there just may be something to the ancient beliefs about salep after all.

It should be noted that there is also an orchid genus *Satyrium*, a name given by Swedish botanist Olof Swartz (1760–1818), one of the first to classify orchids. It consists of more than 80 terrestrial orchids found in Africa, India, Sri Lanka, and

Fig. 1.5 "Saloop, the subject of this etching, has superseded almost every other midnight street refreshment, being a beverage easily made, and a long time considered as a sovereign cure for headache arising from drunkenness. It is a celebrated restorative among the Turks, and with us it stands recommended in consumptions, bilious cholics and all disorders stemming from acrimony in the juices." From *Vagabondiana or Anecdotes of Mendicant Wanderers through the Streets of London*, by artist and engraver John Thomas Smith (London, 1839) (Image courtesy Bishopsgate Institute)

China, with over 30 species from southern Africa. The flowers are non-resupinate (to be explained in Chap. 2, but essentially, the flowers appear to be upside-down) in a variety of colors with the lip having two spurs like a Satyr, which may have been the origin of the genus's name (Fig. 1.9).

Another agreeable orchid-derived drink was Faham tea, made from the fragrant leaves of *Jumellea fragrans* (syn. *Angraecum fragrans*), which grows in mountainous forests on Réunion Island, a dollop of France in the Indian Ocean east of Madagascar. Unlike tea from China, which contained caffeine and thus may induce unwelcome wakefulness, Faham tea was reported to have medicinal properties that were "free from the sleepless effect." Moreover, it had "a most agreeable perfume; after being drunk it leaves a most lasting fragrance in the mouth, and in a closed room the lasting fragrance of it can be recognized long after." [10] Introduced to France, the tea became quite popular in the late 19th and early 20th centuries. However, due to the problems obtaining large quantities of leaves from difficult terrain, the tea was quite expensive, which meant that Faham tea was not really a viable long-term proposition (Fig. 1.10).

Fig. 1.6 This *Phalaenopsis deliciosa*'s enticing pink lip, adorned with yellow with its protruding column and male anther at the top containing packets of pollen, would indeed be thought of as a "prurient apparition" by Ruskin, as would its tasteful name (Image courtesy Alain Brochart)

Returning to Asia, the first full book devoted to orchids comes from China, Chao Shih-Keng's, *Chin Chan Lan Pu*, written in 1233. It described 20 orchid species and their cultivation. Another orchid book followed shortly after in 1247: *Lan Pu*, by Lang Kuei-Lsueh, which described 37 species.

One orchid that has a special place in the culture of the Japanese is the small epiphyte *Neofinetia falcata*. This particular orchid plant has been cultivated in Japan since at least the seventeenth century and is sometimes called the "Samurai Orchid" in reference to the samurai who grew them, although it can be found in China and Korea as well. One way for a feudal lord to curry favor with the 11th shogun Ienari Tokugawa (in office 1787–1837), himself a great admirer of this species, was to bring him an interesting orchid specimen. Even in today's Japanese business culture, giving an orchid gift, particularly a *Phalaenopsis*, is common practice between corporations [11].

Originally, the Samurai Orchid was named *Orchis falcata* by the Swedish naturalist Carl Peter Thunberg, who brought the plant back from his travels in Japan (1775–76). After being part of several different genera over the ensuing years, the genus *Neofinetia* was created in 1925 and named by the notable Chinese taxonomist Hu Xiansu, in honor of the French botanist Achille Finet (1862–1913), who had studied the orchids of Japan and China. Kew now lists this orchid as *Vanda falcata*, with *Neofinetia falcata* as a synonym, whereas on The Plant List, it is just the reverse. Let us persist with *Neofinetia* in honor of Finet. Besides *falcata*, there are now two other recognized species of the *Neofinetia* genus. The flowers have a 3–5 cm nectar spur and a strong, beautiful fragrance, particularly in the evenings, that attracts two species of long-tongue hawkmoths for its pollination [12], as indicated by the table below (Fig. 1.11). The evolutionary aspects of a long nectar spur *vis-à-vis* the length of the pollinator's proboscis, including Darwin's hawkmoth, are discussed in Chap. 4.1.

The cultivation of *Neofinetia falcata* in Japan has a very aesthetic element. The ordinary plants in the wild are termed *fūran*, meaning "wind orchid." However, there are some plants that, due to mutations, exhibit some exceptional property of leaf or flower, and these are the *fūkiran*, meaning "orchid of wealth and nobility." These highly esteemed specimens have a status comparable to that of a noble art form [13]. Recent varieties have appeared with a slightly pink stem and are also highly sought after (Fig. 1.12).

Nineteenth Century Orchid Mania

An interesting thing happened in seventeenth century Holland. The tulip, introduced a few years before the new century began, was becoming enormously popular, especially among members of the new mercantile class. Solid-colored forms and especially variegated forms were highly sought after, and as demand soared, prices did as well, sometimes to astronomical levels for single prized bulb. Futures contracts were developed whereby a buyer and seller would agree on a fixed price for bulbs, to be payable at the end of the tulip season. All manner of speculative practices bloomed, and

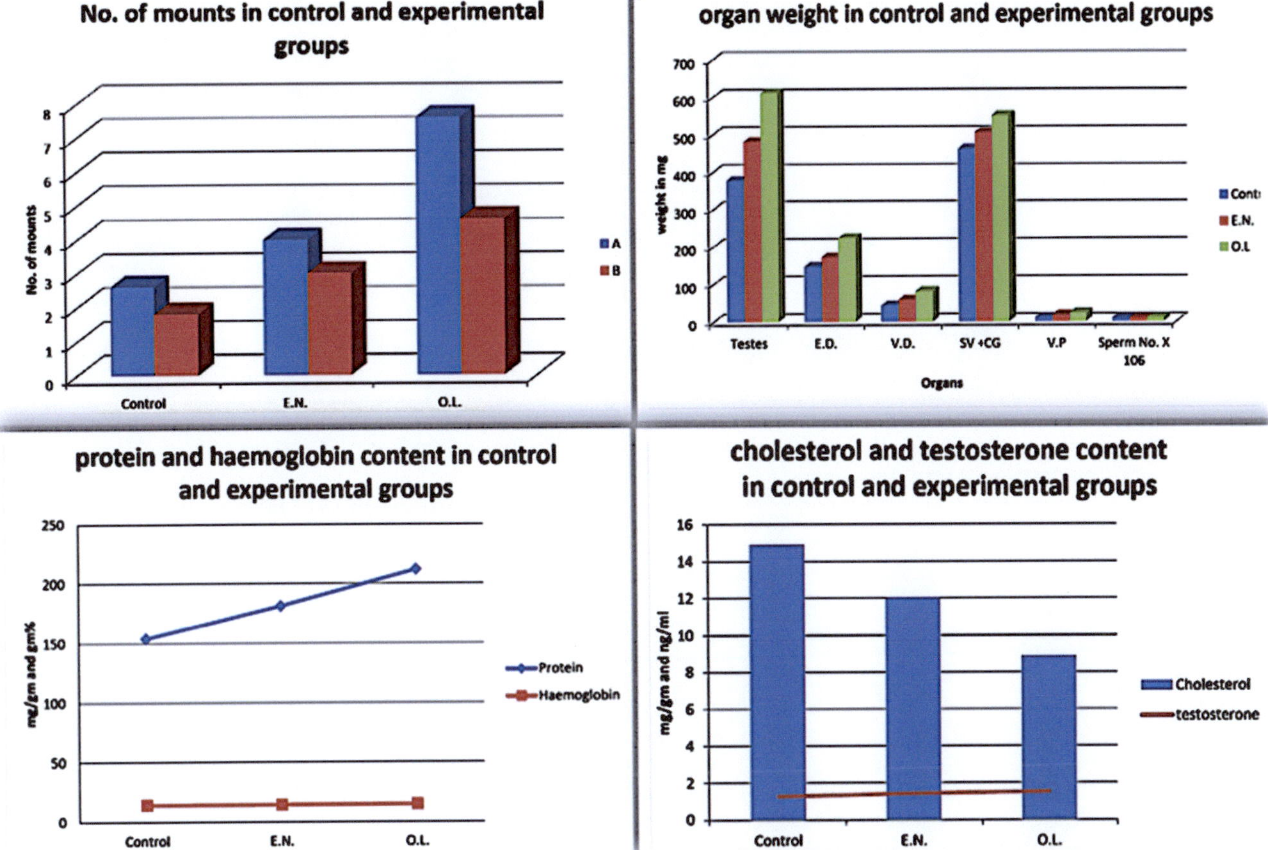

Fig. 1.7 Results from a 21-day study of the medicinal properties of salep reproduced with permission from: [9]. Here E.N. and O.L. represent a powdered feed made up from the ground tubers of the orchids *Eulophia nuda* and *Orchis latifolia* respectively. Six male mice were used in each of the Control, E.N., and O.L. groups. (**A**) – Records the first observation after 15 mins of drug administration; (**B**) – Second observation after 135 mins of drug administration (males and females were separated for 105 mins after first observation). Clockwise: The first chart shows increased mounting behavior for the mice fed *Orchis latifolia*; the second shows increased organ weights of the testes, E.D. = Epidermis, V.D. = Vas deferens, SV + CG = Seminal vesicles with coagulating glands, V.P = Ventral prostrate, including increased sperm count, for the O.L. group; the third chart shows increased levels of testerone (good) as well as decreased levels of cholesterol (good) for the O.L. group; at the scale of the fourth chart the slight increases found in hemoglobin levels (good) are not conspicuous, and there was increased levels of blood protein (neither good nor bad), for the O.L. fed group. Reprinted with permission from [9].

everyone wanted in on the action. Prices escalated further as people began hoarding large quantities of tulips, only to sell them for a profit later on. According to History Professor Anne Goldgar's book on the phenomenon of tulipmania, "People in the 1630s and after found tulipmania a wonder, something to be marveled at, like a fireball, a child with two heads, or a plague of mice." [14].

In February 1637, the economic bubble formed by the tulip market collapsed, providing both historians and economists with a serious matter for study ever since. The specific economic and social intricacies responsible for the collapse are still rather contentious.

A similar fever gripped Victorian England in the nineteenth century over orchids. In the preceding century, England had been introduced to a flowering terrestrial orchid (*Bletia purpurea*), sent from the Bahamas in 1731 (Fig. 1.13). During this period, a few others were being sporadically sent by explorers to Europe and coaxed into blooming. But beyond their scientific study, there was little interest in orchids. All that was about to change.

Fig. 1.8 The romantically enabling *Eulophia nuda (spectabilis)* (Image courtesy Toshiyuki Aoyama)

Fig. 1.9 *Satyrium erectum* from South Africa (Image courtesy Jean-françois Siraudeau)

Fig. 1.10 *Jumellea fragrans* of Réunion island, the leaves of which were made into a deliciously fragrant tea in early twentieth century France. Note the nectar-containing spur at the back of the flower (Image courtesy Frédéric Henze)

In 1817–1818, gentleman and scientist William John Swainson returned from a natural history-collecting expedition in Brazil with a vast array of insects and plants. There are many versions of what happened next. Whether or not Swainson used orchids to "pack his lichens," as horticultural journalist Frederick Boyle had written, is open to question, as Boyle was often unreliable. In any event, William Cattley, a merchant trader, horticulturist, and orchid enthusiast in Barnet near London was in receipt of an orchid specimen sent by Swainson. Cattley flowered the orchid, which was named *Cattleya labiata* in 1824 by the eminent English botanist Dr. John Lindley, thus creating the new genus of *Cattleya*. The *labiata* name

derives from the species' distinctively shaped lip. Lindley later noticed the affinity with the orchid, *Epidendrum violaceum* Lodd., which he renamed *Cattleya loddigesii*. It is interesting to note that *labiata* and *loddigesii* represent the two major forms of *Cattleyas*, the single-leafed unifoliate and the double-leafed bifoliate, respectively. Unfortunately, the location of the discovery of *Cattleya labiata* was lost for more than 70 years (Figs. 1.14 and 1.15).

The feverish period of Victorian Orchid Mania pretty much coincides with the reign of Queen Victoria (1837–1901), and indeed, she was an avid orchid lover herself. Orchids were rare, beautiful, mysterious, and from distant lands, all qualities that had a certain cultural – even snobbish – cachet for the aristocracy, who sometimes paid a fortune to acquire them and create large collections. The Queen herself was honored in 1896 with the naming of *Dendrobium victoria-reginea*, and in more recent times, Chadwick & Sons of Virginia, U.S.A., have taken to naming *Cattleya* hybrids after members of British royalty, including Princess Diana, Kate Middleton, and Queen Elizabeth II, as well as the wives of various US Presidents, among others (Fig. 1.16).

Of course, the mania was not solely confined to the aristocracy and upper classes. It spread among the new wealthy industrialists and mercan-

No.	Photograhed species	order	Frames captured	Times visited	Pollinia attached	Visiting time
1	*Clubiona* sp.	Araneae	1	1	No	23:54
	Drepanopteryx phalaenoides	Neuroptera	8	2	No	1:29-1:41, 2:33
2	*Drepanopteryx phalaenoides*	Neuroptera	1	1	No	20:03
3	*Ceresium* sp.	Coleoptera	1	1	No	20:54
4	*Theretra nessus*	Lepidoptera	1	1	Yes	19:27
5	*Halyomorpha halys*	Hemiptera	6	1	No	22:58-23:00
6	*Ceresium* sp.	Coleoptera	52	1	No	01:07-01: 24
10	*Theretra japonica*	Lepidoptera	1	1	Yes	20:06
12	Mordellidae sp.	Coleoptera	20	1*	No	23:27-23:54*
13	*Polistes* sp.	Hymenoptera	2	1	No	09:53-09:54
17	*Mabra charonialis*	Lepidoptera	221	1	No	22:45-23:52
18	*Mabra charonialis*	Lepidoptera	36	1*	No	00:39-00:52*
20	*Geisha distinctissima*	Hemiptera	34	1	No	20:47-20:55
23	*Mabra charonialis*	Lepidoptera	4	1	No	21:29-21:30

* Observed the floral visitors intermittently with very short interval and thus considered as the same visit.

Fig. 1.11 List of floral visitors captured by interval photography to flowers of *Neofinetia falcata*. Numbers given are the total numbers of frames captured and times visited. The same species captured in consecutive frames were counted as one visit. Among the different insects (spiders, beetles, moths, etc.) only the two species, *Theretra nessus* and *Theretra japonica,* of long-tongued hawkmoth had pollinia attached to them. Adapted with permission from [12].

Fig. 1.12 The delicate *Neofinetia* (*Vanda*) *falcata* is highly esteemed in Japanese culture. Note the slightly pink-tinged, 3–5-cm long nectar spurs. Only two species of long-tongued hawkmoth are known to be the pollinators (Image courtesy Sylvia Kappl)

tilists of the Industrial Revolution. During this time, prices inevitably dropped so that some orchids could be obtained for a few shillings, and thus, the emerging middle classes could benefit from the splendors of orchid growing as well.

The Orchid-Growers Manual, written by Benjamin Samuel Williams (1822–1890), provided much needed expert assistance with orchid cultivation, and it was to become the bible of orchid growers of all classes during this period. The first edition of this work (1852) contained excellent descriptions and notes for the care of "upwards of 260 orchidaceous plants" in cultivation, from *Acineta* to *Zygopetalum*. The thinking was spot-on (except for its male bias): "A knowledge of the different habitats of the various species is essential to the careful grower, so that he may, as far as his means permit, place them in circumstances similar to those in which they make their natural growth." Notes on glasshouse construction, ventilation, growing media for epiphytes and terrestrials, insects, diseases, and propagation were also discussed, including a section not often found in orchid books today: "*Preparing orchids for traveling to a flower-show … Oncidiums* travel well; they require a strong stake to each flower-spike … *Sobralia macrantha* is a bad plant to travel, if not properly tied" (Fig. 1.17). Useful

Fig. 1.13 An illustration of *Bletia purpurea* from *Curtis's Botanical Magazine,* 1833, which first bloomed in England a century earlier

Fig. 1.14 An orchid named by Lindley as *Cattleya loddigesii,* formerly *Epidendrum violaceum* (Image courtesy Norbert Dank/www.flickr.com/photos/nurelias)

advice, even today. Indeed, the *Sobralia cattleya,* having canes up to 6 meters tall, might not be fit for travel at all.

The *Manual* proved so popular that it went through multiple editions over the decades, with the fifth edition (1877) describing the cultivation of upwards of 930 species and varieties. A final sixth edition was published by Williams in 1885, but this was still not enough to meet the demands of the orchid growing public, and Williams' only son Henry published a further edition in 1894 totaling 1050 pages! The Preface to this work states that, *"The Orchid-Grower's Manual* has gained notoriety throughout the civilized world and is even now the text-book of a majority of Orchid Growers." Indeed, by that time, it had even been translated into Russian (Fig. 1.18).

Thus, the mid-to late nineteenth century became a period of great drama regarding the pursuit of new and exotic orchids for the European market. In England, the large commercial empires of Frederick Sander & Co., James Veitch & Sons, Loddiges & Sons, and Messrs. Low, as well as Jean-Jules Linden in Belgium, were in control of dozens of intrepid "travelers" who roamed the most hostile corners of the globe in the search of new and exotic orchids, and the competition was fierce.

The acknowledged "Orchid King" of the period was the German-born but very English Frederick Sander, who was appointed by Queen Victoria as the Royal Orchid Grower. Such was his commercial success that there are countless orchid species that bear his name. At one stage, he commanded 23 travelers in multiple different countries to supply the orchid empire founded on his four-acre nursery at St Albans in Hertfordshire, England. According to orchid authority Merle A. Reinikka, "Kings and nobleman were frequent visitors."

Disease, violent weather, hostile natives and warring tribes, venomous snakes, vicious stinging insects, bloodsucking bats, and wild beasts were daily dangers faced by the courageous orchid hunters who risked their lives for a new species of orchid, with deceit and deception playing an important role in throwing off any other company's travelers who

Fig. 1.15 *Cattleya labiata*, the flower that launched the Victorian Orchid Mania. After its discovery and sensational public appearance, its native habitat was lost for many years (Image courtesy Ramūnas Pileičikas)

Fig. 1.16 Like the woman herself, the stunningly gorgeous *Cattleya* Princess Diana "Wales" (Dubiosa × Hausermann's Gala), named in 2002 (Image courtesy Chadwick & Sons)

Fig. 1.17 The delicate beauty *Sobralia macrantha* is indeed cumbersome to transport to an orchid show, as the flowers appear at the top of thin canes that can extend over a meter tall (Image courtesy Roberta Fox)

might be lurking nearby. This is not to mention outright dirty tricks, which included urinating on the opposition's plants awaiting shipment or paying spies to work for rival firms (Fig. 1.19).

In one 1881 letter from Sander to his traveler Arnold, concerning a beautiful new *Cattleya* he had found, Sander replied: "Keep your gob shut. 1,000 plants would be worth £10,000 if they arrive and are genuine. A fortune! But silence!" [15] In due course, the *Cattleyas* arrived, but they were poorly packed and many had died in transit. In the end, the "new" *Cattleya* was a variety of the lost *Cattleya labiata,* but nonethe-

less brought in a handsome reward for Sander. Indeed, it was only in 1893 that Sander located more of the original *Cattleya labiata vera* (meaning true, not a variety) in the mountains of Pernambuco, Brazil.

All of these daring exploits and intrigue were carried out in order to feed the insatiable European appetite for new and exotic orchids. Prior to this practice, orchid plants had been sporadically collected by botanists. It was in 1840 that James Veitch's nursery sent out the first of many travelers, the young William Lobb, to hunt for more exotic plants to offer the public. And indeed he did, first traveling to South America, then years later to North America, finding a great many new plants, trees, and orchids, one of which is the stunning *Cycnoches pentadactylon* (Fig. 1.20).

Sadly, Lobb suffered a lonely death in a hospital in San Francisco, likely due to syphilis. But his younger brother Thomas, also a collector of orchid species, was sent by Veitch to India and Asia, in turn discovering the dwarf-sized *Phalaenopsis lobbii* (Fig. 1.21).

Fig. 1.18 This wonderful device, called the Thanatophore, was invented by M. Martre of Paris in the 1900's. It was used to fumigate an orchid house by steaming tobacco juice in order to kill such insects as red spiders and thrips. It was advertised in the seventh edition of *The Orchid-Grower's Manual*

A great many other orchid species were discovered by Thomas, including *Aerides fieldingii, Bulbophyllum lobbii (Fig. 1.22), Bulbophyllum reticulatum, Calanthe rosea, Coelogyne flavida, Coelogyne × (Pleione ×) lagenaria, Coelogyne lentiginosa, Coelogyne (Pleione) maculata, Coelogyne schilleriana, Coelogyne speciosa, Cypripedium (Paphiopedilum) villosum*, and these are just the first three letters of the alphabet! [16].

It was assumed at the time that most orchids reaching Europe were from tropical rainforest and jungle regions (as indeed some were), and so they would naturally thrive in very hot, damp, dark glass house environments that were referred to as "stoves." But most did not thrive in such an enclo-

sure, and many of them died as a result. This is because in mountainous habitats where many of these orchids were found, the prevailing temperatures could be quite moderate or even cool. It was the English gardener and architect Joseph Paxton, working for the sixth Duke of Devonshire at Chatsworth, who pioneered the use of separate glasshouse conditions to suit the climatic needs of the different orchid species. He realized the importance of light and air movement in simulating the orchids' natural habitat.

This was the secret to successfully growing orchids in captivity, and Paxton freely published his methods to the benefit of every grower. It was through Paxton that the Duke became entranced with orchids. According to legend, his fascination began upon seeing the *Oncidium* (now *Psychopsis*) *papilio*, and he would go on to own the "finest collection in England" (Fig. 1.29). Unlike most orchids, this particular species can flower on the same inflorescence for many years, so these should not be cut. Orchids that re-bloom on the same inflorescence in this fashion are known as "sequential bloomers." Even *Phalaenopsis* can have this trait (Fig. 1.22).

It was another English horticulturist James Bateman, author of the richly illustrated *The Orchidaceae of Mexico & Guatemala* (1837–1843), who initiated the use of cool-growing conditions for such genera as *Odontoglossum* (Fig. 1.24).

It should be mentioned that Joseph Paxton was also an accomplished architect who designed the famed Crystal Palace in Hyde Park, London, for the 1851 Great Exhibition. It was here that a two-ton specimen of *Grammatophyllum speciosum* was exhibited to dramatic effect (see Chap. 2). Another botanical name for *Grammatophyllum speciosum* is *Grammatophyllum wallisii*, named after the German orchid hunter Gustave Wallis. Wallis was born a deaf mute but somehow managed to speak by the age of six, albeit with a lifelong speech defect. In spite of this disability, he became fluent in several languages, which certainly helped in his foreign travels. He was first employed by Linden to collect plants in South America, where he discovered the gorgeous and fragrant *Cattleya wallisii* (syn. *C. eldorado*) that

Fig. 1.19 Orchid hunter Albert Millican (at right) deep in the Colombian jungle (From *Travels and Adventures of an Orchid Hunter in Colombia*, by Albert Millican, Cassell & Co. Ltd., 1891)

Fig. 1.20 *Cycnoches pentadactylon,* discovered by William Lobb in 1841. The thin curving column indicates that these flowers are males, as the flowers of this genus differ according to sex (Image courtesy Martina Pintaric)

Fig. 1.21 The dwarf-sized *Phalaenopsis lobbii* discovered by Thomas Lobb (Image courtesy Martin Guenther)

grew in the region where the Rio Negro river flows into the Amazon (Fig. 1.25). After years of exploring, Veitch employed Wallis to collect *Phalaenopsis* in the Philippines. His exploring days ended in Ecuador, South America, where he died from the combined effects of disease. He made many new orchid discoveries, and several orchids have been named in his honor, including *Masdevallia wallisii, Odontoglossum wallisii, Neomoorea wallisii,* and *Dracula wallisii* (which he discovered).

Fig. 1.22 A gorgeous discovery by Thomas Lobb, *Bulbophyllum lobbii* (Image courtesy Piotr Markiewicz)

Fig. 1.23 The enticing *"Butterfly Orchid," Oncidium (*now *Psychopsis) papilio,* said to have captured the heart of the bachelor Duke of Devonshire. Who could blame him for succumbing to its charms? (Image courtesy José Pestana)

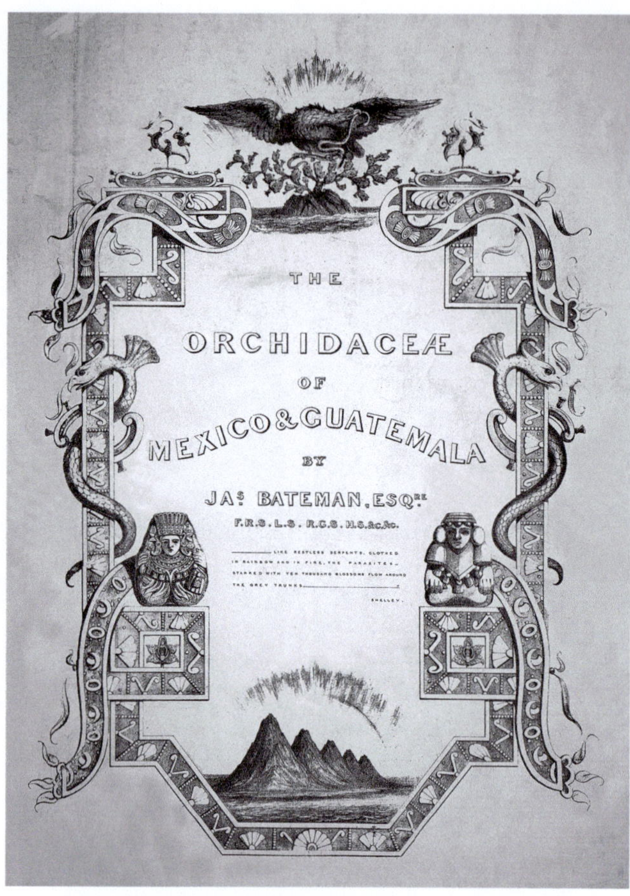

Fig. 1.24 Outrageous in size and price, an original edition of James Bateman's 6-year opus, *The Orchidaceae of Mexico & Guatemala,* was sold in 2002 at Sotheby's auction house for over $100,000 (GBP 55,250)

Fig. 1.25 *Cattleya wallisii,* found by Gustave *Wallis* in Brazil in 1866 (Image courtesy José Amorin)

In the Orchid Mania hothouse environment of the mid-nineteenth century, rumors were rife. One such rumor was that there was a red *Phalaenopsis* growing somewhere in the Philippines. There were of course the white *Phalaenopsis grandiflora* and *amabilis,* as well as the dwarf, *Phal. lobbii,* but the

Fig. 1.26 *Phalaenopsis schilleriana* from the Philippines (Image courtesy Orchi/Wiki Commons)

Fig. 1.27 *Vanda (Euanthe) sanderiana,* only found on the island of Mindanao in the Philippines and accidently discovered by Roebelin after an earthquake in 1880 (Image courtesy Norbert Dank/ www.flickr.com/photos/nurelias)

only truly colored species was the pink *Phalaenopsis schilleriana*, purchased in 1860 for the princely sum of 100 guineas by the Duke of Devonshire (Fig. 1.26). Much searching by commercial orchid hunters turned up nothing, but the red *Phalaenopsis* nevertheless remained highly sought after.

Frederick Sander had read extensively about the southern Philippine island of Mindanao and long held a fascination for it, believing that it was an ideal habitat for as yet unknown species of orchids. After steamship sailings were extended to the island in 1879, Sander dispatched traveler Carl Roebelin to Mindanao in search of the red *Phalaenopsis*.

Roebelin landed in Cotabato in early 1880 and traveled with a guide named Choon around the coast in a small boat to Surigao, at the northern tip of Mindanao. After being told of giant orchids growing on the banks of Lake Magindanao, no sooner had he arrived at the lake than a violent storm came up and he and his guide had to be rescued by local tribesman. Soon thereafter, there came a tense standoff with the tribe involving a trade deal with Choon that had turned sour. Adding to the tension the tribe Roebelin was with became engaged in a pitched battle with a rival tribe! Fortunately for Roebelin, his tribe prevailed in the battle, but further dangers were to come.

The tribal chief provided Roebelin with accommodation in his large tree house. But before dawn, Roebelin was awoken by a mighty roar and violent shaking of the entire forest due to a massive earth-

quake. Bodies were hurled everywhere and the rope ladder to the tree house vanished. When daylight broke, Roebelin found the house was nearly destroyed. Yet, peering through one of the many gaps in the floor, he laid eyes on a beautiful orchid – not the red *Phalaenopsis* but a beautiful *Vanda*, painted mauve, yellow, and brown, subsequently named by the famed German botanist and orchidologist Heinrich Gustav Reichenbach, *Vanda sanderiana* (or *Euanthe sanderiana*, depending on who you consult), and known in the Philippines as Waling-Waling (Fig. 1.27).

The subsequent sale at auction of *V. sanderiana* generated considerable excitement that was typical for that attending the arrival of new or rare orchid species:

> *… The rooms were crowded long before the auction was due to start. The spectacle of hundreds of top hatted orchidophiles arriving by hansom cab or brougham and milling around trying to assess the value of individual plants, the gradual ceasing of chatter while one of the Morrises (of Protheroe and Morris) mounted the rostrum to begin the proceedings in complete silence, must have been an impressive experience [17]*

Indeed.

The pursuit of the red *Phalaenopsis* continued the following year, when Roebelin was sent back to Mindanao for a second time. Sailing along the coast from Davao with his former Chinese guide, he was greeted by a gathering of natives whose

Fig. 1.28 *Phalaenopsis sanderiana,* found on the Philippine island of Mindanao by Carl Roebelin after harrowing losses (Image courtesy Norbert Dank/www.flickr.com/photos/nurelias)

heads were bedecked with red orchids. Once he was taken to the source of the plants, he collected them in their thousands and transported them to Manila to ship back to England. However, disaster then struck:

> *A shameful misfortune has overcome me – destroyed in a few minutes all the plant I had! The lot! In the frightful hurricane which swept across the Philippines on the 28th June (1881) … No question of trying to save the plants – all of them were lost. Every man had to fend for himself [18].*

Some 21,000 plants had been destroyed. So, what does an orchid hunter do in such circumstances? Go back to Mindanao, of course! Roebelin did just that, collecting more of the red *Phalaenopsis,* but in a much smaller quantity than previously. The orchid is not actually red but more a purplish-pink, and it was named *Phalaenopsis sanderiana* by Reichenbach after the Orchid King (Fig. 1.28).

Another notable traveler of Sander's was Wilhelm Micholitz, a German who worked for Sander right up until the First World War. After acquiring thousands of *Dendrobium schroederianum* specimens, a species whose initial habitat had been lost, and not before having to witness a revolting scene of human sacrifices, he reached the island of Celebes (now known as Sulawesi, part of Indonesia) with his consignment. Below is the notorious exchange of telegrams between Micholitz and Sander to illustrate just how serious the orchid business actually was:

SHIP BURNT WHAT DO MICHOLITZ
Sander's reply: RETURN RECOLLECT
Micholitz responds: TOO LATE – RAINY SEASON
RETURN SANDER [19]

And return he did, finding more *Dendrobiums* that were subsequently auctioned in London to much acclaim and profit for Sander.

Yet another Sander collector was Czech horticulturist Benedikt Roezl, who was one of the most prolific orchid hunters of them all, having discovered more than 800 species. He was a tall, handsome fellow with an iron hook for a left hand who roamed throughout South America and Cuba, the western states of the U.S.A., and particularly Mexico. It was in Cuba that he had his fateful machine accident resulting in the loss of his hand, but this only increased his popularity with local natives on his orchid expeditions. He collected tens of thousands of orchid specimens in his 40-year career with Frederick Sander, and his prodigious collecting is one of the reasons for Sander's monumental success.

A fine statue of Roezl was erected after his death in Prague, and he leaves a legacy of several orchids and various other plants and trees that bear his name, including the orchid genus *Roezliella,* which has now been absorbed into the *Oncidium* alliance (Fig. 1.29).

Such was the frenzy and scale of the Victorian Orchid Mania that after the location of *Cypripedium* (*Paphiopedilum*) *spicerianum* was discovered in 1881 by Förstermann in Assam (India), Sander was able to offer 40,000 of the orchid plants for sale at auction on a single day!(Fig. 1.30).

Besides the travelers who worked for established orchid nurseries, other Europeans ventured out on their own in search of rare flowers, such as George Ure Skinner, a merchant trader with business interests in Guatemala. His job allowed him to explore locations far and wide, including Mexico, Central America, and Peru. Over a 30-year period, he began exporting consignments of orchids to England for both study and sale, often taking the orchids with him on his many trips back

Fig. 1.29 Statue in honor of the great orchid hunter Benedikt Roezl in Prague's Charles Square. No hook for a left hand, but holding an orchid instead (Image courtesy Chabe01/Wiki Commons)

to England. His is one of the many names immortalized by orchid species, including *Barkeria skinneri, Cattleya skinneri,* and *Lycaste skinneri.*

The Victorian period of Orchid Mania is filled with interesting characters and daring exploits, and of course, there are far too many to recount them all. Let us mention just two individuals from the century preceding the era of hired travelers, the first being the German-Dutch horticulturist Georg Eberhard Rumphius (1627–1702). Living on the remote island of Ambon off the west coast of New Guinea and initially working for the Dutch East India Company, Rumphius began an intensive study of classifying and describing the flora of the island. One such discovery was a specimen of *Phalaenopsis*, given the name *Angraecum album majus* and described in his monumental catalogue *Herbarium Amboinense* (1661 folio pages and 695 plates). This great testament to the diligence of one man was only published in 1741 and 1750 (the year of publication of the final volume containing the *Phalaenopsis*), many years after his death (Fig. 1.31).

The same *Phalaenopsis* orchid was independently discovered on Java by Pehr Osbeck, a student of Linnaeus, who described it in his *Species Plantarum vol. 2,* as *Epidendrum amabile* in 1753. Little note was taken of *Angraecum album majus* and *Epidendrum amabile* until 1825, when the botanist Karl Ludwig Blume rediscovered the

Fig. 1.30 Not quite numbering 40,000, these *Paphiopedilum haynaldianum* are nonetheless a truly magnificent sight. Found in the Philippines, they are named after Hungarian Lajos Haynald, who had a keen interest in botany and was made a Cardinal in 1879 (Image courtesy Peter Tremain)

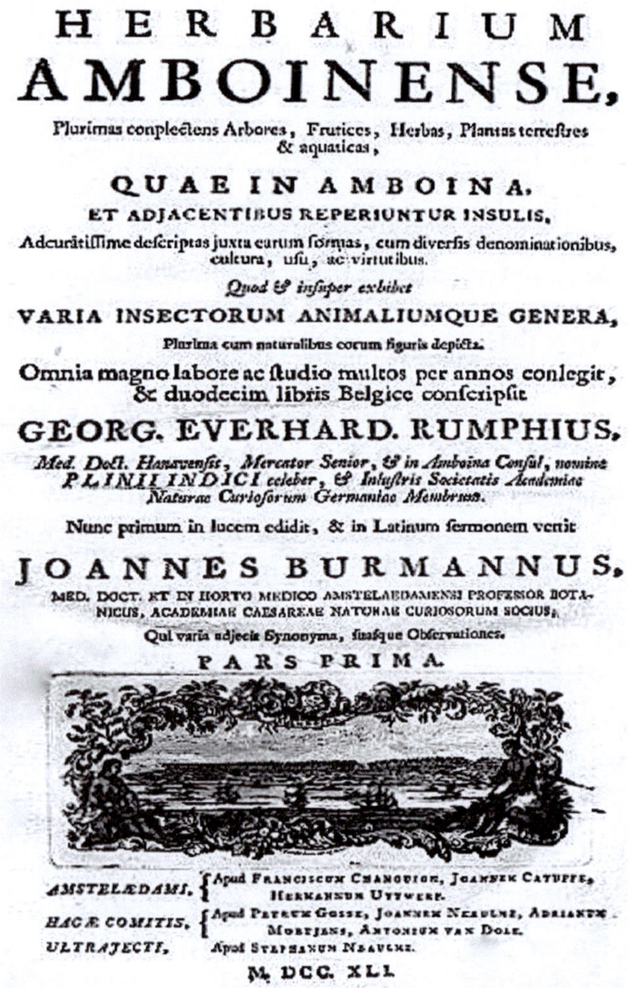

HERBARIUM
AMBOINENSE,
Plurimas conplectens Arbores, Frutices, Herbas, Plantas terrestres & aquaticas,
QUAE IN AMBOINA,
ET ADJACENTIBUS REPERIUNTUR INSULIS,
Adcuratissime descriptas juxta earum formas, cum diversis denominationibus, cultura, usu, ac virtutibus.
Quod & insuper exhibet
VARIA INSECTORUM ANIMALIUMQUE GENERA,
Plurima cum naturalibus eorum figuris depicta.
Omnia magno labore ac studio multos per annos conlegit, & duodecim libris Belgice conscripsit
GEORG. EVERHARD. RUMPHIUS,
Med. Doct. Hanavensis, Mercator Senior, & in Amboina Consul, nomine PLINII INDICI celeber, & Inlustris Societatis Academiae Naturae Curiosorum Germaniae Membrum.
Nunc primum in lucem edidit, & in Latinum sermonem vertit
JOANNES BURMANNUS,
MED. DOCT. ET IN HORTO MEDICO AMSTELAEDAMENSI PROFESSOR BOTANICUS, ACADEMIAE CAESAREAE NATURAE CURIOSORUM SOCIUS,
Qui varia adjecit Synonyma, suasque Observationes.
PARS PRIMA.
AMSTELÆDAMI, Apud FRANCISCUM CHANGUION, JOANNEM CATUFFE, HERMANNUM UYTWERF.
HAGÆ COMITIS, Apud PETRUM GOSSE, JOANNEM NEAULME, ADRIANUM MOETJENS, ANTONIUM VAN DOLE.
ULTRAJECTI, Apud STEPHANUM NEAULME.
M. DCC. XLI.

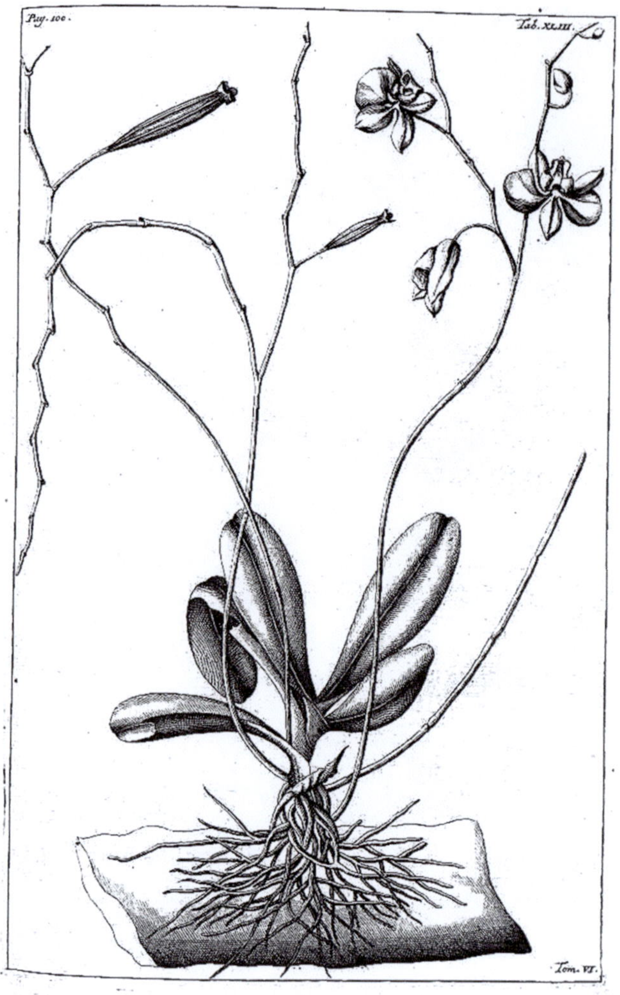

Fig. 1.31 (L): The 1741 title page from Rumphius's monumental catalogue of the flora of Ambon Island. (R): Drawing of the first discovered *Phalaenopsis, Angraecum album majus* from the 1750 *Herbarium Amboinense* by G.E. Rumphius

same species, naming it *Phalaenopsis amabilis* (meaning "charming"). Such has been the success of this one orchid that it now has more than 30,000 registered progeny from hybridization programs (Fig. 1.32).

The island of Ambon itself has been honored through the name of a very fragrant orchid species, *Phalaenopsis amboinensis*, first discovered there. The flower is also found on the neighboring islands of Mulocca, Sulawesi, and Papua New Guinea and grows as an epiphyte. Another orchid bearing the island's name is the *Dendrobium amboinense*, which is endemic to Ambon and nearby islands. The large white flowers exhibit the unusual behavior of opening at night and then shutting for good by the close of the following day.

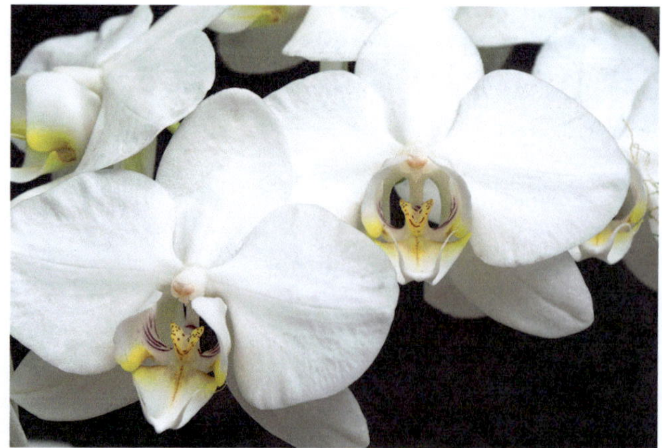

Fig. 1.32 *Phalaenopsis amabilis,* initially discovered and described by Rumphius on the island of Ambon. Discovered again by a student of Linnaeus, and again by Blume, who gave it its current name (Image courtesy Jane Lago)

Fig. 1.33 (L): *Phalaenopsis amboinensis*. (R): *Dendrobium amboinense*, both named for Ambon Island (Images courtesy (L) Martin Guenther, (R) André Fernandez/Cattlaelia)

During this ephemeral process, the flowers transform in color to a pale orange, and the fragrance is modified. Since 2008, this very rare exotic beauty is being grown in captivity (Fig. 1.33).

The second individual is a name that should be known to us all: (Friedrich Wilhelm Heinrich) Alexander von Humboldt (1769–1859), the famous naturalist, explorer, scientist, geographer, and ecologist, who was once considered the "world's greatest living man." Born into a well-connected Prussian family, he chose his own path in life, an intellectual one of science and exploration. Seeking to understand the order of the world, he set out on extensive travels, measuring its every natural and physical parameter. His many popular books provided inspirational reading material for another young naturalist – Charles Darwin – during his own voyage of discovery on the *Beagle*.

After a 5-year exploration of Latin America with French botanist Aimé Bonpland, Humboldt brought back to Europe some 60,000 plants, many not seen before and in need of classification. Humboldt hired German botanist Karl Sigismund Kunth, who assisted with this immense task. Their collaboration resulted in the seven-volume work on new genera and species: *Nova genera et species plantarum quas in peregrinatione ad plagam aequinoctialem orbis novi collegerunt Bonpland et Humboldt* (1815–1825). These volumes were a treasure trove of new botanical discoveries (Fig. 1.34).

Here we find entirely new orchid genera, such as *Odontoglossum, Restrepia, Cyrtochilum,*

Fig. 1.34 Kunth's masterwork describing the thousands of plants brought back from South America by Alexander von Humboldt and Aimé Bonpland

Epistephium, Ionopsis, Trichoceros, and countless new species, such as *Epidendrum ibaguense, Oncidium pictum, Pleurothallis sagittifera* (now *Notylia sagittifera*), *Stelis pusilla,* among many others.

A case of mistaken identity by Kunth was *Cymbidium violaceum,* which subsequently became

Fig. 1.35 (L): The strong, sweetly scented *Zygopetalum maculatum*, brought back from Latin America by Bonpland and von Humboldt and first described by Kunth. (R): The very beautiful *Encyclia cordigera*, known as the "Holy Week Orchid" in Panama, from Central and South America, originally classified by Kunth as a *Cymbidium* (Images courtesy (L) Bernard Dupont, (R) Mabelín Santos)

Cattleya violacea. Kunth wrote in defense, "*pollinis massas in hoc et praecedentibus haud vidi*" – he had not seen the pollen masses in this one or the preceding one, *Cymbidium cordigerum* (now *Encyclia cordigera*). Had he been able to see them, he would have noted that both the *violaceum* and *cordigerum* had four pollen masses, and he certainly knew that *Cymbidiums* had only two pollen masses. Indeed, both *Cattleyas* and *Encyclias* have four pollen masses, and this historically has been used as one anatomical feature separating *Cattleyas* from various other genera that look similar, such as *Laelia,* which has eight pollen masses. However, it must be said that modern DNA evidence has proved this an invalid method of distinguishing genera, and now many species of *Laelia* have been reclassified as *Cattleya* (see Chap. 2) (Fig. 1.35).

A more interesting case was Kunth's misclassification of *Cymbidium candidum*, later to become *Cattleya candida*. Here we find a plea at the end of his description (p.342): "*An hujus generis?*" [20] *Is it this kind?* This raises the possibility that Kunth could have discovered the *Cattleya* genus 8 years prior to it being defined by Lindley, but for some reason he was hesitant to do so.

As with the Dutch tulipmania of the seventeenth century, such abnormal preoccupations of society eventually came to an end. What ended Victorian Orchid Mania was essentially the start of the First World War in 1914. Prices had been softening for years, and in some instances, there was oversupply. Hybrids had been taking over the orchid scene since the turn of the century, as many of the native habits had been plundered bare in order to meet the insatiable demand. No consideration was given to conservation, and gradually, the new and rare species dwindled. With the Great War, Orchid Mania was effectively over.

Despite this setback, orchid popularity has not really died either, as the allure of the orchid is still an incurable condition that afflicts multitudes of aficionados around the world. New orchids are continually being discovered, and in Ecuador alone, a thousand new orchid species have been discovered in the last dozen years. However, very strict regulations are now in place through the international CITES agreement (Convention on International Trade in Endangered Species of Wild Fauna and Flora) to safeguard the survival of the orchids in their native habitat. Severe penalties are

Fig. 1.36 Top: *Oncidium ornithorhynchum,* discovered by von Humboldt and Bonpland and described by Kunth, also known as *Oncidium pyramidale.* The species has been confused with plants of *Oncidium sotoanum* (bottom row) for nearly 200 years, the latter from Mexico and Central America having lavender or pink flowers and given its proper new name in 2010 (Images courtesy (top): Sebastian Moreno, photographed *in situ* at Valle del Cocora, Quindio, Colombia, (bottom R): Alan Cressler, (L) Norbert Dank/ www.flickr.com/photos/ nurelias)

in store for those who breach those regulations. The popular book by Susan Orlean, *The Orchid Thief,* and the subsequent film derived from it, *Adaptation* (2002) starring Nicolas Cage, demonstrate that orchids still have the power to do strange things to people. Moreover, the now-illicit orchid trade, worth millions, remains as ruthless and fraught with intrigue as ever, as narrated in Eric Hansen's *Orchid Fever.*

Indeed, in 2000, English horticulturist and plant hunter Tom Hart Dyke along with traveling companion Paul Winder (a London banker) fell into more than a little trouble while following in the footsteps of their fellow Victorian-era travelers. As they were hunting for rare orchids in the notoriously danger-

ous jungle of the Darién Gap between Panama and Colombia, the two were held hostage for 9 months by a band of suspected FARC guerillas. When the kidnappers' request for a $3 million ransom was refused, the hostages were later released [21].

Outrageous auction prices for particular orchids still crop up from time to time. In 2015, an orchid (Taehwang – emperor, but the species is unclear) sold for a whopping $US 100,000 at a Korean sale. This pales in comparison to the Shenzhen Nongke orchid, which sold in 2005 for roughly $US 202,000, after 8 years of research had created a rather ordinary-looking *Cymbidium* that apparently blooms every 4–5 years but is said to have a rather delicate taste.

Commercial Orchids

Due to their unusual beauty and long-lasting properties, orchids are still in high demand worldwide, both as a cut flower and pot plant. Top exporting nations are the Netherlands, Thailand, Taiwan, and Singapore, and the top export genera are *Phalaenopsis, Dendrobium*, and *Cymbidiums*. Many of the latter bloom during the winter months and add a cheerful aspect to this often gloomy period. Nearly all the orchids commercially sold are modern hybrids, with the large-flowered, cool-growing *Cymbidium* hybrids bred from just a handful of native species. These are very popular plants to grow outdoors in temperate climates.

During the 1940s and 50s, it was *de rigueur* for a woman going out on a formal occasion to wear a *Cattleya* corsage. As various large-flowered *Cattleyas* bloom at different times of the year, this orchid could meet the demand of any social occasion.

The genus *Phalaenopsis* of Moth Orchids (*phalaina* is Greek for "moth," and *opsis*, "like") has now become a favorite with the general public. Whereas *Cymbidiums* would grow outside in a sheltered position, the modern *Phalaenopsis* hybrids like it a bit warmer and prefer the internal coziness of the house. Indeed, in any magazine featuring a modern house interior, it is almost an essential feature to have an orchid reposing unostentatiously on a tabletop. A *Phalaenopsis* remains a statement of refinement and sophistication, a statement of class. And this class, bequeathed by an exotic-looking *Phalaenopsis,* is now available at many florists and even supermarkets at a very modest price. Most *Phalaenopsis* have no scent, but one species that does is the lovely *Phalaenopsis bellina* from Borneo and Malaysia (Fig. 1.37).

Another unusual fragrant *Phalaenopsis* species is *Phal. tetraspis*, native to the Andaman and Nicobar Islands and parts of Sumatra. Normally, the flowers have from one to several bars running across some of the sepals or petals in purple or deep red. However, there is an unusual variety, *C#1*, which has one or more sepals or petals often entirely red (Fig. 1.38).

Although this book is not a grower's guide, given its modern popularity, it is worth mentioning

Fig. 1.37 The wonderfully scented Phalaenopsis bellina from Borneo and Malaysia (Image courtesy Azhar Ismail)

the commercial prescription for getting *Phalaenopsis* to flower: Spiking can be initiated through exposure to a daytime temperature of 25 °C (77 °F) and nighttime temperature of 20 °C (68 °F) for a period of 4–5 weeks. The cool temperature then needs to be maintained, or the shoot will turn into vegetative growth. Once the *Phalaenopsis* has flowered, unlike most other orchids, it can be coaxed into flowering again by cutting the inflorescence about 1 centimeter above a triangular node that lies just beneath the lowest flower. There – it has been said.

Much of the current *Phalaenopsis* mania is due to modern hybridization. Making orchid hybrids has a long history, going right back to the period of Victorian Orchid Mania, when in 1856, John Dominy, working for Veitch's Nursery, flowered the cross he made 3 years earlier, *Calanthe furcata × Calanthe masuca*, to produce the hybrid *Calanthe* Dominyi. Since then, over 539,000 orchid hybrids have been registered (as of April 2017 according to *OrchidWiz*), not only within the same species but between different genera as well.

Let us examine a simple hybrid case. We will start with the pollen from *Phalaenopsis inscriptio-sinensis,* a recent discovery found on the Indonesian island of Sumatra and described in 1983. The pol-

Fig. 1.38 A typical *Phalaenopsis tetraspis* (top) and the *Phalaenopsis tetraspis C#1* with a random red petal (Images courtesy Martin Guenther (top), and Francis J. Quesada Pallares (bottom))

len will be used to fertilize an old acquaintance, the beautiful white *Phalaenopsis amabilis* (Fig. 1.32). The hybrid primary cross is known as *Phalaenopsis* Little Spot and was carried out by Luc Vincent in 2006 (Fig. 1.39).

In general, a modern day hybrid can have a very large and complex family tree with both species and other hybrids coming into the act. One example that is not too complex would be *Phalaenopsis* Cleopatra, registered in 1974. Each hybrid in the ancestors would have its own family tree, so the chart is only part of the ancestry story [22] (Figs. 1.40 and 1.41).

And what does the hybrid orchid look like? Simply gorgeous (Figs. 1.42).

We should not forget, however, that Mother Nature herself is a master hybridizer, as many natural orchid crosses occur in the wild. Their highly elaborate and colorful markings are to be marveled at, although they are not really there for our benefit! (Fig. 1.43)

Vanilla

Unless you are an orchid enthusiast, you might not realize that vanilla comes from an orchid plant that grows in the wild in parts of South America, Central America, and Mexico. Studies have indicated that the genus goes back 60–70 million

Fig. 1.39 (L): The markings on Sumatran *Phalaenopsis inscriptiosinensis* look like Chinese characters. Its pollen was used to fertilize the white *Phalaenopsis amabilis* in order to create the delightfully spotted hybrid cross, *Phalaenopsis* Little Spot (R) (Images courtesy (L) Martin Guenther, (R) Luc Vincent)

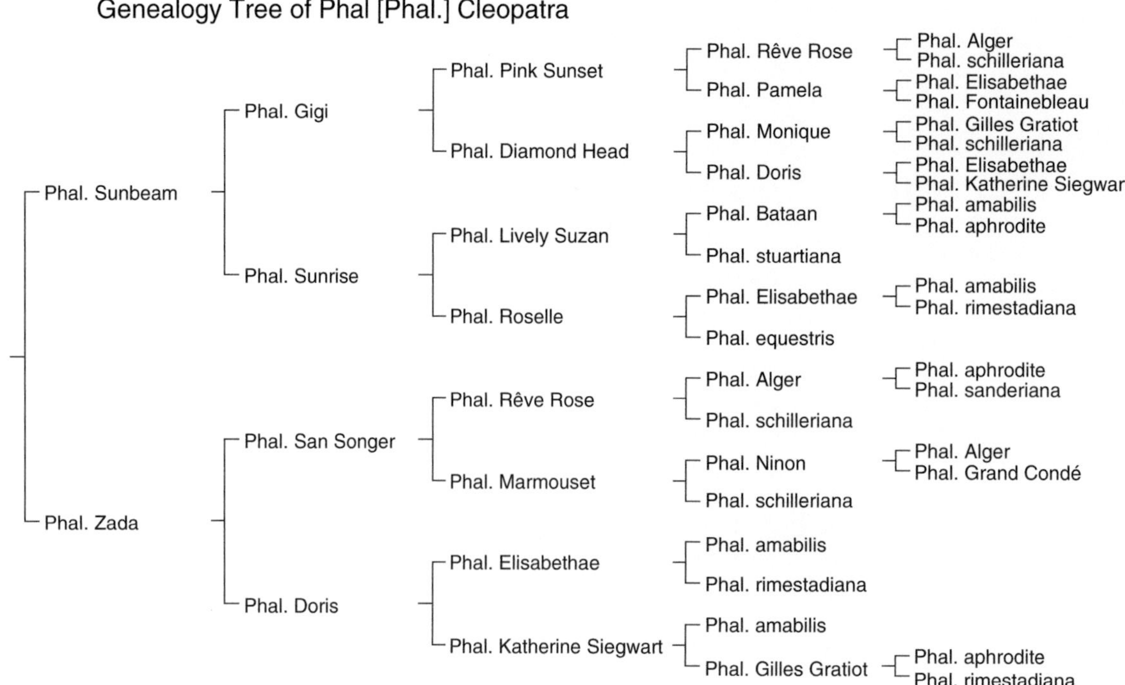

Fig. 1.40 The ancestry of hybrid *Phalaenopsis* Cleopatra, going back to its roots (Image courtesy OrchidWiz)

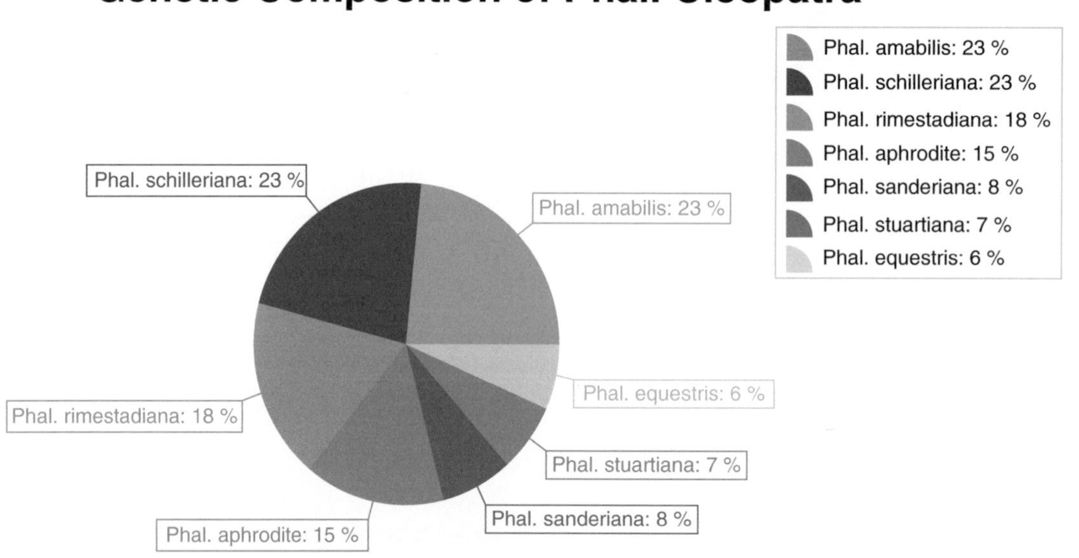

Fig. 1.41 The complete parentage of *Phalaenopsis* Cleopatra (Image courtesy OrchidWiz)

years. Although there are more than 100 different species, nearly all of the natural vanilla flavoring used in the world comes from a single species, *Vanilla planifolia* (also known as *Vanilla fragrans*). It is a creeping perennial vine that climbs up trees and prefers partial shade and warm, humid climates. Often, it is grown under an existing tree canopy that provides the necessary sun protection.

It was the Spanish explorer Hernán Cortés who witnessed the drinking of chocolātl made from cocoa and flavored with vanilla in the Aztec court of Montezuma (Moctezuma II). The great chief was said to drink up to 50 cups a day. Actually, the

Fig. 1.42 *Phalaenopsis* Cleopatra (Image courtesy Boris O. Schlumpberger, Landeshauptstadt Hannover Fachbereich Herrenhäuser Gärten)

Fig. 1.43 The Ecuadorian *Odontoglossum × elegans* (now *Oncidium × bockemuhliae* according to Kew) is a natural cross between *Odm. cristatum* and *Odm. cirrhosum* (Image courtesy Steve Beckendorf)

Fig. 1.44 The historical medicinal uses of vanilla (Adapted from [29])

Medical uses of vanilla

Author	Year	Proposed use of vanilla derivatives
Aztec herbal[24]	1552	Flavouring and perfume prevent fatigue in those holding public office. Bestow the bodily strength of a gladiator
		Drive weariness far away
		Drive out fear and fortify the human heart
Menashian *et al.*[25]	1992	Improve food intake and reduce nausea and vomiting in patients given chemotherapy
Fladby *et al.*[26]	2004	Diagnostic of Alzheimer's disease (patients cannot smell vanilla)
Fitzgerald *et al.*[27]	2004	Antimicrobial against *Eschericha coli*, *Lactobacillus planatarum* and *Listeria innocua*

vanilla was discovered by the Totonacan people, who gave it as an offering to their Aztec conquerors. Cortés shipped the vanilla back to Spain in 1519, and it was subsequently cultivated by botanical gardens in France and England. Surprisingly, the flower has only a very faint sweet scent and does not smell like vanilla.

By the eighteenth century, vanilla was being recommended as a tonic to increase sexual desire. *King's American Dispensatory* (1898) promised that vanilla would, "stimulate the sexual propensities" and was "considered an aphrodisiac, powerfully exciting the generative system." Interestingly, such a notion still persists to the present day. "When consumed, vanilla improves sexual debilitation and kicks the central nervous system into gear, causing sensations to feel even better," asserts a 2017 website for a hotter sex life [23]. Certainly, no one can argue with the feel-good factor of the vanilla fragrance and taste (Fig. 1.44).

For commercial production, the flowers must be pollinated by hand early in the morning, as the flowers only last one day. An efficient method of doing so – using a stick and thumb gesture – was discovered by a 12-year-old slave named Edmond Albius on the island of Réunion near Madagascar, where the plant had been introduced by the French. The method is still in use to this day. The green seed pods can grow up to 20 cm (8″) in length and are ready for harvesting by hand roughly 6–9 months after pollination. The harvested pods are cured, generally by placing them into hot water for a few minutes, and then dried in the sun and aged.

Actual commercial practice is very labor intensive as the drying pods need to be brought inside every night, which is why the resulting dark brown liquid is so expensive. The vital fragrant ingredient that develops in the cured pods is the aromatic compound, vanillin ($C_8H_8O_3$) (Fig. 1.45).

Vanilla has become one of the world's most popular flavorings. It has found its way into cosmetics, confectionary, chocolate, and all manner of culinary delights and beverages. What a poor world it would be without the fruits of this outstanding orchid! Yet, less than 1% of vanilla flavoring is derived naturally, with the rest being made artificially, mostly from the

Fig. 1.45 Top (L): The orchid flower *Vanilla planifola*. (R): Row upon row of vanilla plants grow on upright posts at this vanilla plantation on the island of Réunion. Bottom: Vanilla seed pods ripening at a plantation in Sulawesi, Indonesia (Images courtesy Top (L) Giancarlo Sibilio, Naples Botanical Garden, University of Naples Federico II, Italy, (R) David Monniaux/ Wiki Commons; Bottom: Manfred Sommer)

organic compounds guaiacol and lignin. There are some 18,000 products on the global market that use vanilla flavoring, [28] and the supply of natural vanilla simply cannot meet with this demand.

Indonesia, Madagascar, Mexico, Papua New Guinea, and China are the top five leading producers of natural vanilla (2016). It is a highly valuable commodity ranking second only to saffron as the world's most expensive flavoring. For some purposes, the whole, dried pods (of which 2% is vanillin) are used as an infusion, while for others, such as certain baking recipes, only the seeds are required, as in such heavenly desserts as *crème brûlée*.

Other varieties are grown on the islands of Tahiti and Moorea, such as *Vanilla tahitensis,* which has slightly different characteristics and is mainly used in ice cream. Additionally, *Vanilla pompon* from the islands of Guadeloupe and Martinique is used in perfume and pharmaceuticals.

References

1. S.R. Ramírez *et al.*, Dating the origin of the Orchidaceae from a fossil orchid with its pollinator, *Nature*, 448, 1042–1045, 2007.
2. Charles Darwin, *The Various Contrivances by Which Orchids are Fertilized by Insects and On The Good Effects of Intercrossing*, John Murray, London, 1862, 1st ed., p.307.
3. T.J. Gavnish, *et al.*, Orchid historical biogeography, diversification, Antarctica and the paradox of orchid dispersal, *J. Biogragr*, 1-12, 2016.
4. D. Chamovitz, *What a Plant Knows, A Field Guide to the Senses*, Scientific American, 2012.
5. P. Calvo and K. Friston, Predicting green: really radical (plant) predictive processing, *J. Roy. Soc. Interface*, 14, 1–20, 2017.
6. A. Trewavas, Aspects of Plant Intelligence, *Ann. Bot.* 92(1), 1–20, 2003; A. Trewavas, *Plant Behavior and Intelligence*, Oxford University Press, 2014.
7. http://www.rjwhelan.co.nz/herbs%20A-Z/eyebright.html.
8. M. Grieve, *A Modern Herbal*, Jonathan Cape Ltd, 1931, reprinted 1998, p.605.
9. S.P. Jagdale, *et. al.*, Pharmacological Studies of 'Salep', *Herbal Medicine and Toxicology* 3(1), 153–156, 2009.
10. Quoted in: *Medicinal Orchids of Asia*, by Eng Soon Teoh, Springer, 2016, taken from a flyer in the 1920s.
11. Article in *Japan Times*, July 21, 2017: http://www.japantimes.co.jp/news/2017/02/02/business/japans-business-gift-culture-says-orchids/
12. There is strong evidence for this assertion: K. Suetsugu *et al.*, Potential pollinator of *Vanda falcata* (Orchidaceae): *Theretra* (Lepidoptera: Sphingidae) hawkmoths are visitors of long spurred orchid, *Eur. J. Entomol.* 112(2), 393–397, 2015.
13. For those with more than a passing interest in this beautiful orchid, there is the All Japan Fūkiran Society and its offshoot: Fūkiran Society of America: http://fukiransoa.weebly.com/f363kiran-books.html
14. Anne Goldgar, *Tulipmania: Money, Honor, and Knowledge in the Dutch Golden Age*, University of Chicago Press, 2007.
15. Arthur Swinson, *Frederick Sander, The Orchid King*, Hodder & Stoughton, 1970, quoted from p.54.
16. Taken from *Hortus Veitchii*, by James H. Veitch, 1906, that documents all the orchids discovered at Veitch's Nursery and by whom.
17. Peter McKenzie Black, *Orchids*, Hamlyn, 1973, p.71.
18. Swinson, p.70.
19. Swinson, p.105.
20. *Kunth in F.W.H.von Humboldt, A.J.A.Bonpland & C.S.Kunth*, Vol. 1, 1815, p.342. This issue is also addressed in: Ruben P. Sauleda, The proper name for a Colombian *Cattleya* Lindl., *New World Orchidaceae – Nomenclatural Notes*, No. 6, 2013.
21. Tom Hart Dyke and Paul Winder, *The Cloud Garden – A True Story of Adventure, Survival, and Extreme Horticulture*, Lyons Press, 2004.
22. Hybrid parentage as well as genera and species information can also be found on the BlueNanta website: http://bluenanta.com/
23. http://stylecaster.com/12-aphrodisiac-foods-to-eat-for-hotter-sex/#ixzz4cwvRPfRs
24. Reinikka M.A., *A History of the Orchid*, Timber Press, Portland, OR, 1995.
25. L. Menashian, *et al.*, Improved food intake and reduced nausea and vomiting in patients given a restricted diet while receiving cisplatin chemotherapy. *J. Am. Diet. Assoc.* 92, 58–61, 1992.
26. T. Fladby, *et al.* Olfactory response in the temporal cortex of the elderly measured with near infrared spectroscopy: a preliminary feasibility study, *J. Cereb. Blood Flow Metab.*, 24, 677–80, 2004.
27. D.J. Fitzgerald, *et al.*, A. Mode of antimicrobial action of vanillin against Escherichia coli, Lactobacillus plantarum and Listeria innocua. *J. Appl. Microbiol.*, 97, 104–13, 2004.
28. As quoted in: M.M. Bomgardner, The Problem with Vanilla, *Chem. & Eng. News*, Sept. 2016.
29. C.J. Bulpitt's "The uses and misuses of orchids in medicine," *Q.J. Med* 98, 625–631, 2005.

"They have come twelve thousand miles to look at a flower?" Bati asked me in Malay.
"It is true," I replied.
"Can you eat this flower?" Katong asked.
"No."
"Is it used for medicine?"
"No."
"What do they want to do with this flower?"
"Take photographs and measure the leaves."

Eric Hansen in *Orchid Fever*

Orchidaceae is the largest family of flowering plants in the world, having conquered every continent except Antarctica. As of 2017, the number of genera is somewhere between 736 and 900, with 28,000 species, but this number is increasing every year [1]. This is out of approximately 300,000 flowering plant species in total. About 14,000 of the orchid species are epiphytes, in that they grow on trees, although they take nothing from the tree except a perch for support. Epiphytes are mostly found in Mexico, Central and South America, as well as Asia and the Pacific Islands. Terrestrial orchids are mostly found in North America, Europe, Australia, and New Zealand.

The orchid-blessed countries of Colombia and Ecuador have over 4000 orchid species each, with Costa Rica being honored with the greatest density of orchid species [2].

Databases of the currently recognized orchid genera and species are given by the Kew Royal Botanic Garden website, The Plant List, Tropicos (Missouri Botanical Gardens), and BlueNanta, given in the Bibliography.

Taxonomy

So much diversity, but how to make sense of it all? By the middle of the 18th century, orchids had been trickling into Europe for a few hundred years from explorers of Africa, the Americas, Asia, and the Pacific.

It was Joseph Pitton de Tournefort (1656–1708), a French botanist, who was the first to distinguished orchids as a separate family from other plants in his *Élémens de Botanique* (1694), classifying a few orchid genera including *Orchis,* which received its official botanical blessing.

A great breakthrough in plant taxonomy (and subsequently the taxonomy of all living things), came from Carl Linnaeus (1707–1778) who in his 1753 seminal work *Species Plantarum* introduced the binomial system that is still utilized today. Prior to this time, and even in a preceding monograph by Linnaeus on orchids, *Species orchidum et affinium plantarum* (1740), a polynomial system was employed to classify plants. This involved a Latin (genus) name, followed by a cumbersome description also in Latin of the plant itself. Not only was the name difficult to remember, but different botanists would use different descriptions (Fig. 2.1).

The *Species Plantarum,* employing a binomial system stating the genus first, then the species, contained eight orchid genera (*Cypripedium, Epidendrum, Herminium, Neottia, Ophrys, Orchis, Satyrium, Serapias* – all of which are still valid genera), and 62 species of mostly European origin. In all, Linnaeus' two-volume work classified 5900 plants in this fashion, which rose to over 9000

© Springer International Publishing AG 2018
J.L. Schiff, *Rare and Exotic Orchids*, https://doi.org/10.1007/978-3-319-70034-2_2

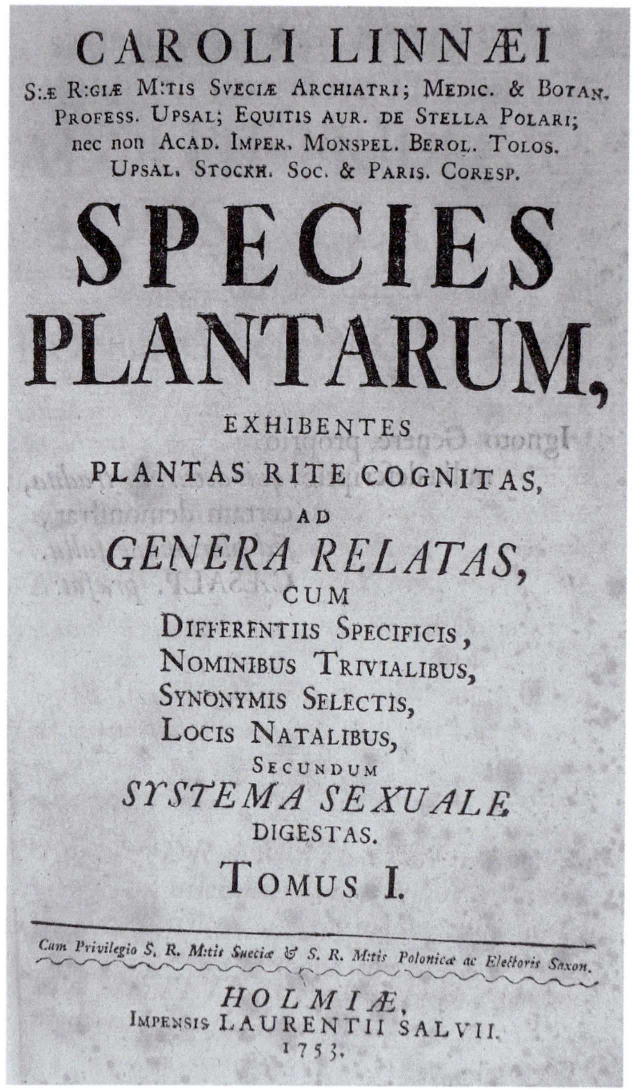

Fig. 2.1 The modern classification of all plants began with the binomial system introduced by Linnaeus in 1753

Fig. 2.2 The green *Pabstia viridis* from Brazil (Image courtesy Orchi/Creative Commons)

identified plants in his subsequent work. The *International Code of Botanical Nomenclature* (ICBN), the modern set of regulations for the naming of plants, still uses the binomial system invented by Linnaeus. It has its origins in the International Botanical Congress of 1867, an entity that still exists today and continues to revise the original ICBN. Its most recent meeting convened in 2011, where it was decided to include algae and fungi on this list [3].

Orchid names are often abbreviated, and a list of abbreviations is included in Appendix VI.

An orchid species is a group of plants with the same general appearance and characteristics, such as the same flower, leaves, and pseudobulb (if any) morphology, same growth and blooming habit, and structure of the sexual apparatus. Of course, they must also be capable of interbreeding. More recently, the same genetic structure has been added to this list of required traits. As Nature is not a Swiss watch, some variation is allowed in size and color, and some species members can be epiphytic while others are lithophytic or terrestrial etc. Often the species epithet is named for someone in the field of botany, an orchid grower, or its discoverer, as we have seen in Chap. 1, or it sometimes has to do with the appearance of the plant flower itself, such as *elegans, fimbriata* (fringed), *fragrans, floribunda, grandiflora, viridis* (green), or its habitat location, such as *tahitensis,* or, *sinense* (Fig. 2.2).

The genus is a grouping of different species that all share similar characteristics. Over the past quarter millennium, hundreds of new genera (the plural of genus) have been created, (and some subsumed into other genera like the former genus, *Schomburgkia*), in order to better discriminate and understand the diverse range of orchids in the natural world. This has resulted in many species being shuffled from one genus to another, yielding a range of historical *synonyms*.

As an example, the lovely, compact *Promenaea stapelioides*, endemic to Brazil, was named by Dr. John Lindley in 1843 when he created the genus *Promenaea* by removing several species from the existing genus, *Maxillaria* (Fig. 2.3). Lindley's

Fig. 2.3 The patterned *Promenaea stapelioides* from Brazil (Image courtesy Eric Hunt)

name, *Promenaea stapelioides,* is that by which it is known today, but it has several historical synonyms:

Cymbidium stapelioides (1821)
Maxillaria stapelioides (1832)
Peristeria stapelioides (1850)
Zygopetalum stapelioides (1863)

Notice that two of the synonyms occur after Lindley's 1843 designation. Many other orchids have had an even more varied history as botanists have struggled to make sense some 28,000 species of orchids. Modern DNA studies are bringing further refinements not solely based on plant morphology. As a consequence, there are either 736, 765, 856, 899, etc. orchid genera depending on which source you consult [4]. And of course, these numbers will change over time.

Additionally, a number of species have varieties that differ in some small degree from the type species due to mutations or genetic variation in the same way that one's children look somewhat different from one another. The popular and easily grown *Laelia anceps* from Mexico and Honduras has many varieties (Fig. 2.4).

There are various competing taxonomic philosophies concerning what should be the most defining criteria, and the subject has now become very arcane and fraught with controversy. Entire books have been written on the subject. On this complex matter, let Dr. Mario A. Blanco, Director of the

Lankester Botanical Gardens of the University of Costa Rica, have the last word: "Classifications have never been truly stable. And this has to do with the fact that every classification is artificial. There is no such thing as a 'natural' classification … Individual plants occur in nature; classifications do not." [5].

Thus, for the most part, we will adopt the more traditional nomenclature, because that is what most orchid aficionados are comfortable with, including this author. For example, orchid lovers know well the beautiful *Laelia purpurata,* the national flower of Brazil. But recent DNA evidence has moved it along with various other Brazilian *Laelias* into the *Cattleya* genus, and it is to be known henceforth as *Cattleya purpurata.* Like *Laelia anceps*, it also comes in many varieties (Fig. 2.5).

Two other taxonomic terms are useful when it comes to species (and hybrids): *cultivar* and *grex.* A cultivar is a particularly outstanding example of a species or hybrid that could merit an award, such as *Laelia anceps* "Sand Bar Pink Parade." All genetically identical plants arising from either cloning or keikis from *Laelia anceps* "Sand Bar Pink Parade" will also bear the same cultivar name. A grex is the second name given to a manmade cross, such as *Cattleya* Michelle Obama, which is a cross between the species *Cattleya trianaei* and another hybrid *Cattleya* Mini Purple). All other crosses between the same orchid pair will bear the same name, *Cattleya* Michelle Obama, although, like children, they will not necessarily look exactly the same (Fig. 2.6).

A few orchid genera have more than a thousand different species, such as *Bulbophyllum* (over 2000), *Epidendrum* (over 1500), and *Dendrobium* (over 1500), while some have a mere single species member. One such is *Neomoorea* and its stunning sole species member *Neomoorea wallisii,* which grows in the wet conditions of the cloud forests in Colombia, Ecuador, and Panama. The genus is named after the former curator of the Glasnevin Botanic Garden in Dublin, Frederick W. Moore, and the species after the famous German orchid collector Gustave Wallis, whom we encountered in Chap. 1. In the past, the orchid has gone by the

Fig. 2.4 Clockwise from left: *Laelia anceps; L. anceps var. alba; L. anceps var. guererro; L. anceps var. veitchiana* (All images courtesy Roberta Fox)

Fig. 2.5 *Laelia purpurata*, now officially known as *Cattleya purpurata* as a result of modern DNA evidence (Image courtesy Peter Tremain)

names *Lueddemannia wallisii* and *Moorea irrorata*, with the present name dating from 1924. The flowers are 6–7 cm across and fragrant, with dif-ferent plants exhibiting somewhat variable markings on the labellum (Fig. 2.7).

For those of us who will never visit a cloud forest with its persistent low clouds and abundant moisture, below is one in Costa Rica. In such an environment, many species of orchid are yet to be discovered (Fig. 2.8).

Anatomy

Orchid flowers are distinctly different from conventional flowers. For one thing, orchids have a bilateral symmetry where the left side is the mirror image of the right side. Moreover, most orchid flowers are hermaphroditic with the male and female sexual organs of the orchid flower found on a single *column*, replacing the separated pistil and stamen of other flowers. As the sexual organs are in close proximity, they are

Fig. 2.6 Top: The cultivar *Laelia anceps* "San Bar Pink Parade." Bottom: The hybrid *Cattleya* Michelle Obama (Images courtesy (top) Van Swearingen; (bottom) Chadwick & Sons)

often separated by a ridged membrane known as the *rostellum* that separates the female *stigma* from the male *anther*. The pollen is also different, so that in orchids, all the grains are bound together into coherent masses (pollinia), whereas in ordinary flowers, the pollen remains powdery.

Orchids have developed many floral strategies to promote cross-pollination, and the rostellum has evolved to inhibit self-pollination, although the latter does occur due to various factors, including scarcity of pollinators. (See Chap. 4). And indeed, some species like *Cypripedium* do not have a complete rostellum or one that is underdeveloped, as in *Cephalanthera*. Charles Darwin was very interested in the rostellum from an evolutionary point of view, for, as he observed, "No organ like the rostellum exists in any other flower." [6].

Fig. 2.7 The butterfly-like *Neomoorea wallisii* (Image courtesy Ruud de Block)

Orchid flowers do a very strange thing. For the majority of orchids, when the blooms are still in the bud stage, the lip is the topmost petal. But as the flower begins to open, the stem twists around 180° to present the bloom with the lip at the bottom. The flower is called *resupinate*, which means it is actually upside down, although this is the way most people view the flower and consider it to be right-side up. Some genera do not go through this rotation (*Ponthieva*, some *Prosthechea*, some *Satyrium*, *Scaphosepalum*, etc.) and are known as *non-resupinate*. These flowers look upside-down, but are not. And strangest of all, the flowers of *Angraecum superbum* from Madagascar rotate from the bud state through a complete 360°, so that they return to their original non-resupinate position! (Fig. 2.9).

Flower Terminology

Osmophores: tissues on a flower that produce and secrete the volatile substances responsible for flower's scent.

Anther cap (*Anther*): the part of the column bearing the pollen masses that are housed inside its cap-like structure.

Column (*Gynostemium*): a single columnar structure containing both the male and female parts of an orchid.

Fig. 2.8 The cloud forest in Costa Rica provides abundant moisture for many orchid species, known and unknown (Image courtesy Mabelín Santos)

Fig. 2.9 The flowers of *Angraecum eburneum subsp. superbum* (*Angraecum superbum*) rotate a full 360° from bud to fully open bloom (Image courtesy Kevin Holcomb)

Pollinium (pl. *pollinia*) a coherent mass of pollen grains, unlike other flowers, where the pollen grains are a fine powder. In general, orchid flowers have either 2, 4, 6, or 8 pollinia. Pollinia come in two basic types: soft and mealy or hard and waxy.

Stipe and *caudicle*: the stipe is a stalk-like structure connecting pollinia to the viscidium, which is variable in length and thickness. Caudicles are thin strands that attach pollinia to each other or to the viscidium.

Viscidium: a viscid disk that attaches to a pollinator. Some orchids do not have a viscidium, but rather the base of the caudicle has an adhesive substance exuded from the rostellum that attaches to the pollinator.

Pollinarium (pl. Pollinaria): The whole structure of pollinia, stipe, and viscidium, forming the pollination unit.

There is considerable variation in the form and size of the pollinarium as there is with most orchid features. In Fig. 2.11 is a sample from Southern Brazil.

Stigma: the female part of the flower, generally a shallow depression, that receives the pollen and located on the column.

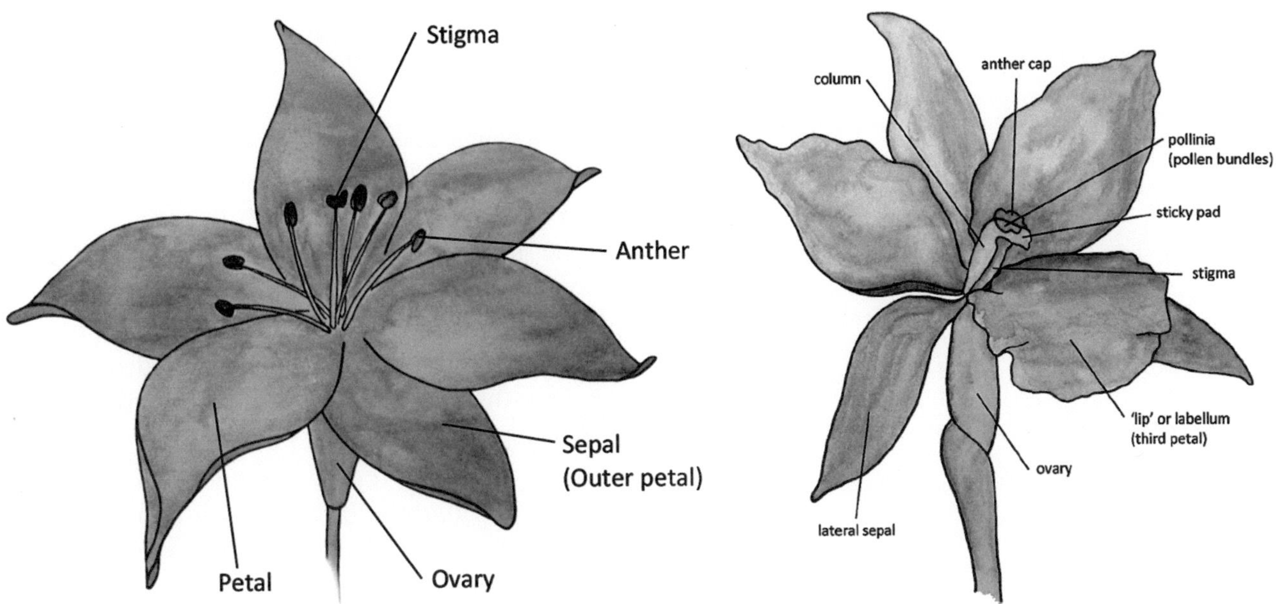

Fig. 2.10 Orchid flowers (right) are unlike all others in that they have distinct structural differences, particularly in regards to their sexual organs (Illustration by Katy Metcalf)

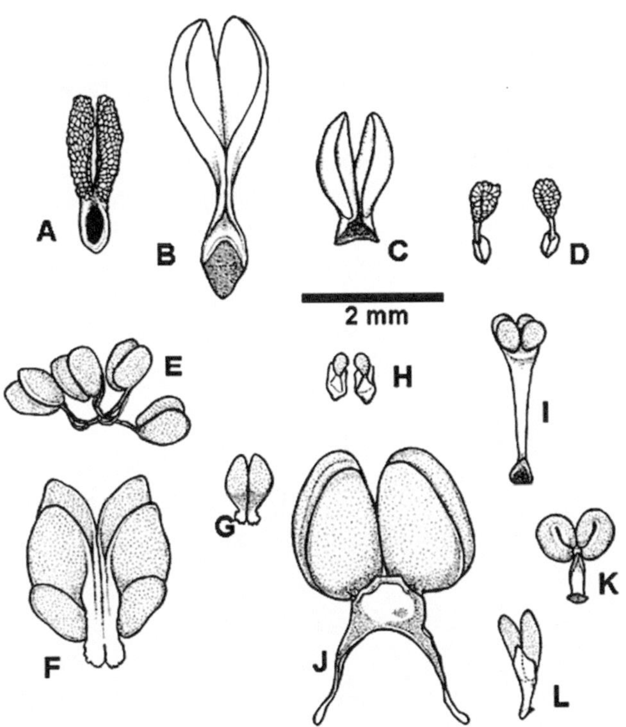

Fig. 2.11 (**a**). *Microchilus austrobrasiliensis* (**b**). *Veyretia simplex* (**c**). *Cyclopogon diversifolius* (**d**). *Habenaria parviflora* (**e**). *Isabelia pulchella* (**f**). *Leptotes unicolor* (**g**). *Acianthera luteola* (**h**). *Campylocentrum aromaticum* (**i**). *Zygostates dasyrrhiza* (**j**). *Brasiliorchis picta* (**k**). *Oncidium loefgrenii* (**l**). *Oncidium paranaense* (Image courtesy R.B. Singer *et al.* [7])

Rostellum: an outgrowth that separates the pollinia from the stigma and inhibits self-pollination (Fig. 2.12).

Perhaps Darwin's fascination with the role of the rostellum was due to the contradictory fact that he married his first cousin, Emma Wedgood, yet was certainly well aware of the robustness gained through genetic diversity. Indeed, the subtitle of his book on *Contrivances* (see Bibliography) was "On the Good Effects of Intercrossing." In the end, after drawing up a list of pros and cons of marrying Emma, Darwin concluded: "Marry — Mary — Marry Q.E.D."

Even more remarkable is how the rostellum results in the flower achieving cross-fertilization. Firstly, the flower is   by a pollinating insect expecting to get some kind of reward for its efforts. It may be nectar, it may be sex, and the many and varied contrivances by which the orchid lures the insect to perform its duty will be investigated in Chap. 4 and 5. Then, the insect will find the pollinarium attached to some part of its body (e.g. head or thorax), as the orchid flower has struck the first blow and the glue from the viscidium begins to set.

Darwin noted that in the case of *Orchis mascula*, the stipe will then bend into a horizontal

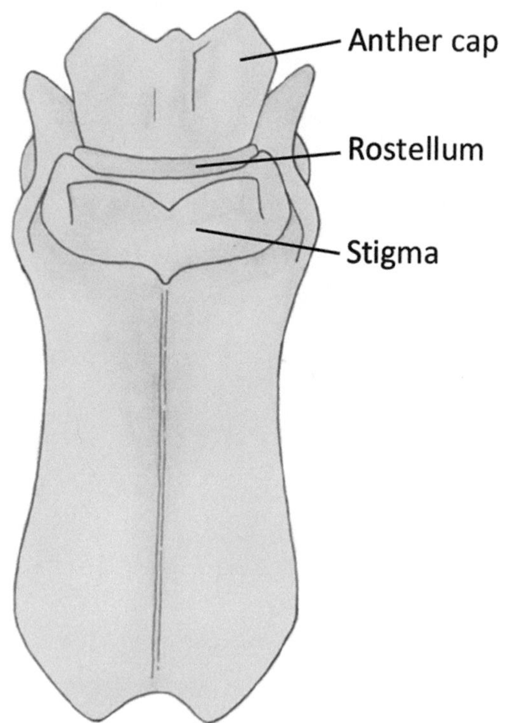

Fig. 2.12 The anther, rostellum, and stigma on the column (Illustration by Katy Metcalf)

position towards the front of the insect so that when the insect visits another flower of the same species, the pollinium will not push up against its former position but rather brush against the stigma lower down on the column. The stigmatic surface is sticky, and as a consequence, the pollinium breaks away from the stipe, completing the pollination process (Fig. 2.13).

On the leaves, there are very important structures known as *stomata,* which are minute apertures that open and close to allow the passage of carbon dioxide in and oxygen and water out during photosynthesis. Their role will be discussed further in the next section (Fig. 2.14).

Pseudobulbs

One distinctive feature of many orchids is the characteristic appearance of pseudobulbs, thickened upright stems that branch off the rhizome of sympodial epiphytic orchids. Terrestrial orchids with plentiful moisture and nutrients have little need for pseudobulbs, but the existence of an epiphyte is more precarious, often characterized by only intermittent rainfall or periods of drought. Studies have

shown that when subjected to drought stress, the water and chlorophyll content of the pseudobulbs was significantly more reduced than in the leaves, thereby considerably mitigating the drought effects on the leaves. This capability is vital for the survival of the plant [8].

The name "pseudobulbs" distinguishes them from true bulbs that are found in other terrestrial plants such as daffodils, lilies, tulips, onions, etc. The orchid pseudobulb is a thickened portion of the stem that stores water, minerals, carbohydrates and other nutrients for the plant's vegetative development, including reproduction (Fig. 2.15).

But the pseudobulb does much more. The initiation of the inflorescences, which generally emanate from the base or the top of the pseudobulb as well as the flowering process of the orchid, is mediated by the organic chemistry within the pseudobulb. Pseudobulbs can vary both in size and shape depending on the genus, from small and spherical, large and pleated, to flattened and oval. Some are shaped like thickened canes, as in many species of *Dendrobium,* which also has species with more conventionally shaped bulbs. Members of the *Catasetum* genus have a dry, dormant period, where the leaves turn yellow and are dropped, but once rain water becomes available again, the pseudobulbs spring awake with new life from the stored nutrients.

Photosynthesis and CAM

In normal plant photosynthesis, chlorophyll in the leaves transforms sunlight into chemical energy that is used to synthesize sugars from carbon dioxide (CO_2) absorbed from the air through leaf stomata and from water taken up by the roots. Oxygen (O_2) is vented as a waste by-product out through the stomata. The stomata close at night or during hot, arid conditions to prevent water loss.

However, there are many orchid genera (as well as other plants including cacti) that grow in somewhat arid conditions or experience irregular rainfall. As a consequence, they have evolved a different version of the photosynthesis process from other plants, called CAM (Crassulacean

Fig. 2.13 (L): A pollinium of *Orchis mascula* as it would initially attach to an insect and the position it subsequently assumes for fertilizing the stigma. (L) Illustration by Katy Metcalf, (R) *Orchis mascula* (Image courtesy Olivier Debré)

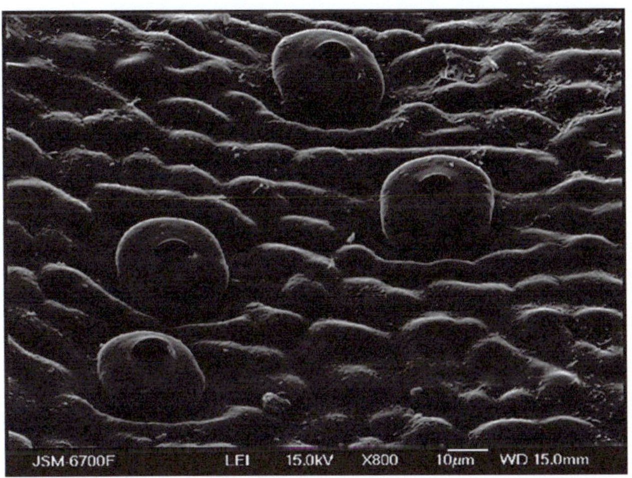

Fig. 2.14 Stomata found on the leaf of *Cattleya gaskelliana* (Image courtesy Lyudmyla Buyun)

Acid Metabolism). "CAM has been found in 62% and 26% of epiphytic orchid species in Australian and New Guinean rainforests, respectively, 42% of orchid species in a moist lowland forest site in Panama, and up to 100% of the epiphytic flora in a Mexican dry forest. The abundance of CAM species in such habitats is related to limited water availability ... CAM species are also found in contrasting habitats such as very arid and very moist sites." [9] Indeed, "orchids have more CAM species than any other family" and can be found growing on cacti and succulents [10].

In CAM, which is the reverse of ordinary photosynthesis, the stomata in the leaves close during the day in order to avoid losing water vapor through transpiration, but then during the night when it is cooler, the stomata reopen in order to acquire CO_2. However, the process still requires sunlight, so the CO_2 is stored overnight in the form of malic acid ($C_4H_6O_5$). During daylight, the malic acid is converted back to CO_2, and conventional photosynthesis is initiated. The "Crassulacean" in CAM derives from the *Crassulaceae* family of mainly succulent plants in which this metabolic process was first studied, and the "Acid" comes from the plant's utilization of malic acid.

Growth Habitats

Most orchids have *sympodial* growth habit in that they have a rhizome along which new vegetative growths form. Others, such as *Sarcochilus, Vanda,* and *Vanilla,* are *monopodial* and grow along a single stem (Fig. 2.16).

Fig. 2.15 Clockwise: Square-sided pseudobulbs (*Bifrenaria harrisoniae*); ovoid pseudobulbs (*Bulbophyllum nutans*); orchid cane pseudobulbs (*Dendrobium moniliforme*); flat-tened ovate, spotted pseudobulbs (*Maxillariella vulcanica*) (Images courtesy (clockwise): Dalton Holland Baptista, Rémi Tournebize, Don Brown, Naoki Takabayashi)

Sarcochilus are very floriferous, compact plants found in Australia and some Pacific islands. They resemble miniature *Vandas* and are either epiphytes or lithophytes, forming clumps as they grow. Many species have a delicate red circular pattern surrounding the central region like a target, and some, such as *Sarco. falcatus,* are fragrant (Fig. 2.17).

Terrestrial Orchids

Terrestrial orchids grow in a variety of habitats from open meadows, forest floors, and swampy fields, and even arid, sandy conditions. Some orchids, such as the ones of the *Orchis* genus, produce tubers beneath the ground, while others like *Cypripediums* produce only roots along a creeping rhizome. Some terrestrial orchids have pseudobulbs, such as *Phaius*, and some are deciduous, such as *Calanthe*, which also has species that are evergreen (Fig. 2.18).

On the other hand, some orchids such as *Crepidium acuminatum* (common name *Malaxis acuminata*), found in India and parts of Asia, have tuber-like pseudobulbs growing in terrestrial and lithophytic environments. Interestingly, an extract from the dried pseudobulbs is used in

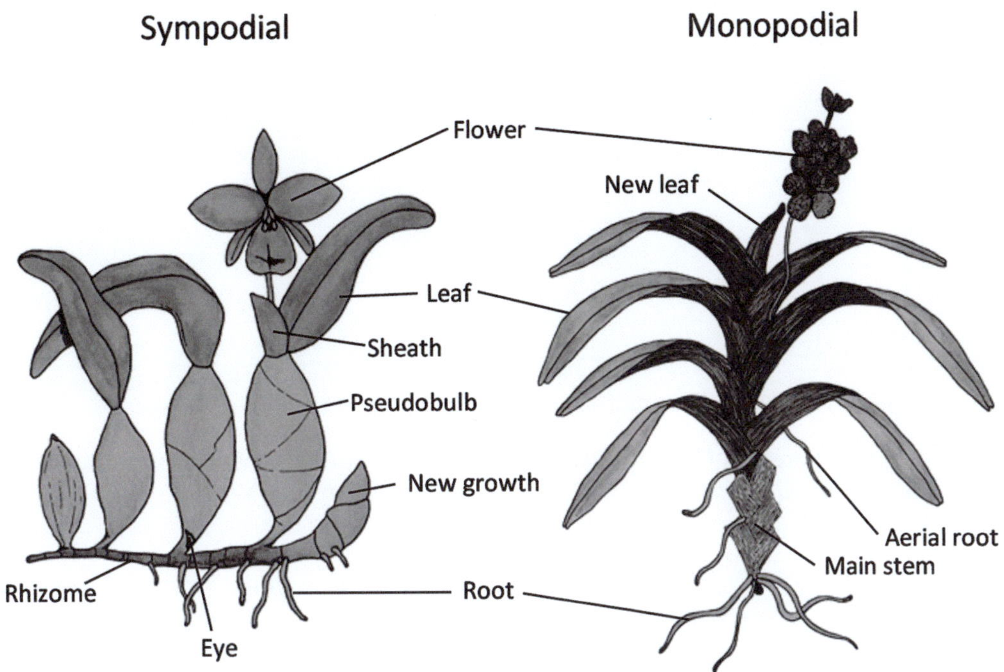

Fig. 2.16 The two basic growth habits of orchids (Illustration by Katy Metcalf)

Fig. 2.17 (L): This miniature species of *Sarcochilus falcatus* is an Australian native with the common name "Orange Blossom Orchid," owing to its fragrance. (R): The delicately patterned hybrid *Sarcochilus* Fitzhart, both of whose parents (*Sarco. hartmanii* × *Sarco. fitzgeraldii*) are native to Australia (Images courtesy Kath Knight)

Ayurveda as a tonic to treat fever, tuberculosis, and general debility, as well as to stimulate the production of semen. Recent scientific research has found the pseudobulbs contain antioxidants including polyphenols, [11] which are important nutrients for human health. In this same study, the authors demonstrated that phytochemicals derived from an extract of the pseudobulb could be used to create nanoparticles of gold at room temperature! (Fig. 2.19).

Fig. 2.18 (L): The evergreen terrestrial *Calanthe reflexa,* found in Southeast Asia and the Himalayas. (R): The terrestrial deciduous orchid *Bletilla ochracea* from South-Central China (Images courtesy Tom Velardi)

Fig. 2.19 The tuberous-like pseudobulbs beneath the flowers of *Crepidium acuminatum* are used in Ayurvedic medicine to treat a whole host of symptoms (Image courtesy Alok Mahendroo)

And some, such as the bizarre *Rhizanthella*, grow completely underground, as will be seen in the next chapter.

Epiphytic Orchids

It is estimated that approximately 72% of all orchids (by number, not species) are epiphytic. They are found mostly in tropical regions growing perched on the lower branches of trees or high up in the tree canopy. Nothing is extracted from the tree except hospitality, with the orchid's roots gripping the bark firmly. The roots absorb moisture and nutrients from the air and from what washes off the branches, such as bird droppings and the

like. The tree also provides dappled sunlight, which is favorable to most epiphytes (Fig. 2.20)

Lithophytic Orchids

Lithophytic orchids mainly grow on rocks in tropical habitats. Their roots grip the rough surface and can burrow into the cracks. A few lithophytic species are not too fussy and can also grow as epiphytes or terrestrials or even in all three environments, such as *Thunia alba* from China, India, and Myanmar (Fig. 2.21).

Roots

The roots of epiphytic and lithophytic orchids are distinct from those of terrestrial orchids, and indeed other terrestrial plants, in that they grow exposed to the air and light and as such engage in photosynthesis. Epiphytic and lithophytic roots have multiple layers of a soft material known as *velamen*, often a gray-white color, which absorbs water and nutrients for the plant and protects the root cortex from drying out. The green growing tips of the orchid roots are evidence of chlorophyll (Fig. 2.22).

Recent research has indicated that cyanobacteria found in the velamen of orchid roots is involved with nitrogen fixation. "Since [an] insufficient amount of minerals is common in the tropics, nitrogen may be the limiting factor for growth and development of epiphytic orchids. Involvement of various cyanobacteria in formation of the sheath-shaped covering suggests that nitrogen fixation is

Fig. 2.20 The epiphyte *Ansellia africana,* found in tropical and southern Africa (Image courtesy Carlos Cruz/flickr/ epiphyte78)

the primary function of orchid-associated cyano-bacteria." [12] It is also suggested that the cyano-bacteria provide nutrients to the plant's associated mycorrhizal fungi.

Root-colonizing bacteria also play an important role in the root's development and in overall plant health. Interestingly, many of these bacteria have antifungal properties, and many of the roots' associated orchid fungi have antibacterial properties. Somehow, the host orchid manages to keep all these competing factions in just the correct balance.

Birds and the Bees

Living organisms have to reproduce in order to survive, and orchids are faced with their own set of challenges to attract a suitable pollinator. Thus, a lot of their world revolves around the subject of sex. According to authors Van der Pilj and Dodson (1966), the various orchid pollinators are: listed in (Fig. 2.23) [13].

Hummingbirds pollinate a number of orchids when they attempt to gather nectar, including most species of *Elleanthus,* *Cyrtochilum retusum,* *Stenorrhyncos speciosum, Arpophyllum giganteum,* and a few species of *Sobralia.* The latter have two thickened ridges at the base of the lip known as *calli,* which secrete nectar. Typically, the pollen grains become stuck to the hummingbird's beak, and in many cases, the pollinia happen to be dark in color, whereas most orchid pollinia are yellow. It has been suggested that the reason for this dark color adaptation is so that the hummingbird does not attempt to clean off the pollinia, which would not be favorable for the orchid [14] (Fig. 2.24).

Seed Production

The seeds of an orchid are the smallest in the world of flowering plants, with many about the size of a particle of dust. The very smallest orchid seeds are 0.085 mm, just beyond the resolution of the

Fig. 2.21 (Top) The lithophytic *Paphiopedilum stonei* growing on a limestone cliff face in Sarawak, Malaysia. (Bottom): *Thunia alba (marshalliana)* grows as an epiphyte, lithophyte, or as a terrestrial and is found at high altitudes (1000–2300 m) in China, India, and the Himalayas (Image courtesy (top) Chien C. Lee, (bottom) Daniel McLaren)

unaided eye of 0.1 mm. As a consequence, seed pods can carry from thousands to several million seeds per pod. Darwin himself counted 6200 seeds from one pod of *Orchis maculata* (and 20 times that number of pollen grains); one pod of a species of *Acropera* was reported to contain 371,250 seeds; *Epidendrum radicans* can have up to a half-million seeds per pod; and a *Maxillaria* seed pod was found to contain 1,756,440 seeds [15].

Darwin made a back-of-the-envelope calculation based on *Orchis maculata* (now *Dactylorhiza maculata*), and found that if nearly all its seeds from all its pods led to viable plants, the great-grandchildren of a single plant would nearly cover

the entire land area of the Earth! A pleasing thought, but obviously this does not happen, so the question is why not? (Fig. 2.25).

The main reason why the Earth is not knee-deep in orchids is that the seeds require a *mycorrhizal* fungus in order to even germinate. The seeds of most other flowering plants encapsulate the embryo with *endosperm* tissue containing vital nutrients of starches and proteins for the embryo's initial phase of development. Humans consume endosperm all the time whenever they eat grains, rice (indeed, white rice *is* endosperm), or coconut (Fig. 2.26).

On the other hand, an orchid seed, which is often microscopic in size, has no endosperm, and so it must enter into a parasitic relationship (called *myco-heterotrophy*) with a mycorrhizal fungus [16] in order to germinate the embryo contained within. Some orchid seeds require a very specific mycorrhizal fungus; others are less particular about which one provides the required sugars and other nutrients necessary for the embryo to develop. Most adult orchids retain their mycorrhizal partnerships, and due to their characteristically poorly developed root systems, they are thought to be heavily reliant on mycorrhizal fungi for mineral nutrition [17]. This is one reason for the great number of orchid seeds produced in a single pod. Once the pod ripens and splits open, casting its contents to the four winds, the seeds must find just the appropriate conditions under which to germinate. Survival of the species would seem to be a numbers game, but not entirely. As observed by Darwin:

> *The number of the individuals which come to maturity does not seem to be at all closely determined by the number of seeds which each species produces; and this holds when closely related forms are compared. Thus Ophrys apifera fertilizes itself and every flower produces a capsule; but the individuals of this species are not so numerous in some parts of England as those of O. muscifera, which cannot fertilize itself and is imperfectly fertilized by insects, so that a large proportion of the flowers drop off unimpregnated [18].*

Darwin then goes on to give several other similar examples from other parts of the world (Fig. 2.27).

Another reason the Earth is not covered in orchids is because of the few numbers of flowers that are successfully pollinated and develop a seed

Fig. 2.22 A jungle of roots of aerially suspended Vanda orchids above a crowd at Kew Gardens, London. The *Vandas* will quite happily grow in this fashion (Image courtesy Howard Carshalton)

Wasps and Bees	60%
Moths	8%
Butterflies	3%
Flies	15%
TOTAL INSECTS	86%
Birds	3%
Mixed Agents	8%
Autogamous	3%
TOTAL	100%

Fig. 2.23 Here, "Autogamous" refers to self-pollination. The "Mixed Agents" category includes bats and frogs, and the "Insects" category includes ants, beetles, and now even a raspy cricket (see Chap. 4.1)

Fig. 2.24 A hummingbird feasts on the nectar of *Arpophyllum giganteum,* found in the forests of the New World. Orchid grower Roberta Fox says the flowers are a "hummingbird magnet." (Image courtesy Roberta Fox)

pod. In fact, a flower can be pollinated without a subsequent seed pod developing.

Propagation Techniques

Natural Propagation

A very simple method of natural propagation is to divide the existing plant into clumps. This is trivial for terrestrial orchids. For epiphytes with pseudo-bulbs, the rhizome should be cut so that each division has a new growth or bud from which to grow. However, this is not always possible, and if there are some remaining backbulbs, that is, pseudobulbs that do not have a new growth, then these can also be made to sprout from a dormant bud. This can be done by placing the backbulbs in

Fig. 2.25 Not quite covering the Earth, but still bountifully flowered *Orchis* (*Dactylorhiza*) *maculata* in a swampy field near St Petersburg, Russia (Image courtesy Alexey Sergeev)

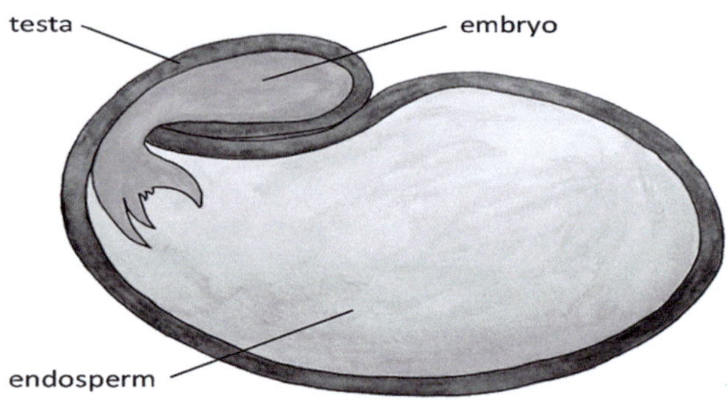

Fig. 2.26 A typical non-orchid seed contains endosperm that provides nutrients for the development of the embryo. This is a crucial difference with orchid seeds, which have no endosperm (Illustration by Katy Metcalf)

sphagnum moss and keeping them warm and damp until new growth is initiated.

It should be mentioned that sphagnum has been used for dressing wounds for hundreds of years due to its highly absorbent and antiseptic properties. It was used extensively during World War I for this purpose and is superior to cotton dressings. It makes an excellent growing medium for *some* orchids as it retains moisture far longer than other media.

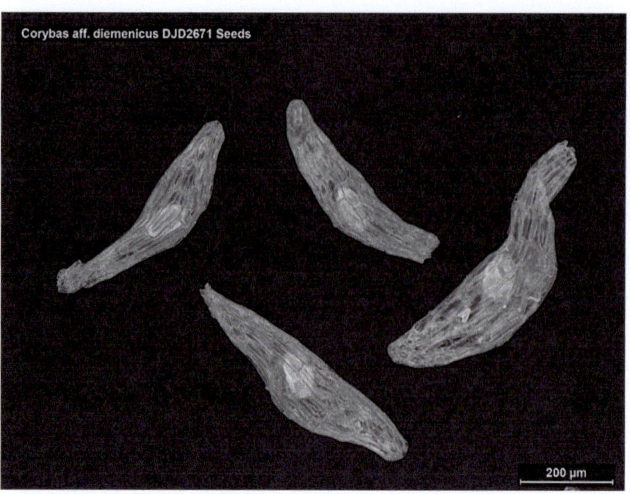

Fig. 2.27 Seeds from an endangered species of *Corybas* from Victoria, Australia. Lengths are about 0.4 mm, with orchid seeds among the world's smallest (Image courtesy South Australia Seed Conservation Centre)

Other natural propagation methods include cutting up 15 cm (6″) lengths of *Dendrobium* canes, or of *Phalaenopsis* stems (which contain dormant buds of course), and laying them down in a tray of damp sphagnum moss. This can be similarly done for lengths of the flowering stems of *Phaius tankervilleae*, even if laid prone in moist soil [19]. Within a few weeks, plantlets will start to develop.

Phaius tankervilleae, known as the "Swamp Orchid" or "Nun's Orchid," is a stunning terrestrial orchid possessing pseudobulbs and found widely spread throughout Asia and Pacific Islands, including Australia, Africa and Madagascar. The species was named by famed botanist Sir Joseph Banks in honor of Lady Emma Tankerville (1752–1836), a collector of exotic plants who flowered the orchid from a specimen brought back from China. It is actually on the endangered species list due to the usual suspects associated with human encroachment. Thus, home propagation is especially desirable (Fig. 2.28).

There are some orchids, however, that produce their own offshoot plantlet known as a *keiki* (meaning "baby" or "child" in Hawaiian), which develops its own root system while on a stem (e.g. *Epidendrum radicans*) or soft-caned *Dendrobiums*. Even a *Phalaenopsis* will sometimes produce a keiki. Once the keiki has sufficient roots, it can be planted out (Fig. 2.29).

Fig. 2.28 *Phaius tankervilleae* (Image courtesy Nova de Jong)

Fig. 2.29 Several keikis growing on one cane of a *Dendrobium loddigesii* (Image courtesy Carlos Cruz/flickr/epiphyte78)

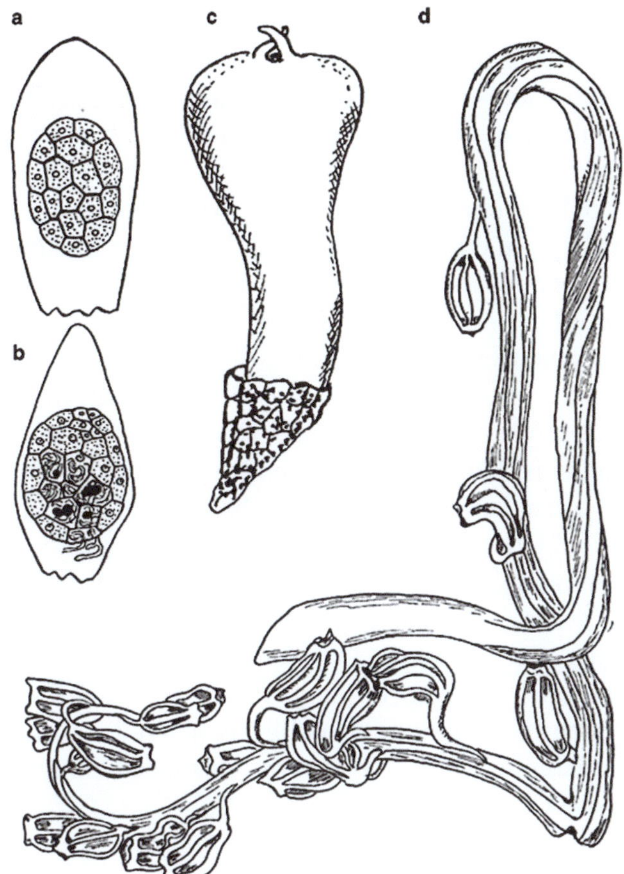

Fig. 2.30 Noël Bernard's serendipitous discovery of a dead inflorescence of *Neottia nidus-avis* buried in soil. He found within the seed pods all of the above stages: (**a**) undeveloped seed, (**b**) seed beginning to germinate after being invaded by a fungus, (**c**) formation of a protocorms, (**d**) dead inflorescence (From Bernard 1902) [21])

Artificial Propagation

Since orchids reproduce with some difficulty in their natural habitat, and they are protected by the CITES Treaty, scientists have developed various laboratory techniques in order to satisfy world demand for these wonderfully exotic plants.

Victorian orchid breeders interested in producing hybrids would sow the seed around the base of the mother plant on the potting media. This method had a very low rate of success. When the first *Phalaenopsis* cross was made in 1875 at Veitch's Nursery between *Phal. amabilis* and *Phal. equestris* by the same method, only a single plant survived to subsequently flower. But the success rate improved when in 1899, French botanist Noël

Bernard "while walking in the Fontainebleau forest near Melun… discovered a dead, broken inflorescence of the achlorophyllous orchid *Neottia nidus-avis* that suggested to him a theory for orchid seed germination." [20].

Indeed, Bernard realized that it was a soil fungus that provided all the water, minerals, and carbon nutrients for the embryo to germinate. He subsequently identified fungi of the genus *Rhizoctonia* from the roots of various orchids and was able to get their seeds to germinate in the lab. Tragically, Bernard's brilliant career was cut short at age 37 by tuberculosis (Fig. 2.30).

However, the above method had its problems, and the orchid seeds often failed to germinate or resulted in the death of the embryo. A different approach to the culture of orchid seeds *in vitro* was worked out in 1922 by the American plant biologist Lewis Knudson, working at Cornell University [22]. Knudson realized that the mycorrhizal fungi could be dispensed with if the orchid embryo was provided with the same nutrients that the fungus provides (asymbiotic culture). He formulated a growing medium, "Knudson B," that was a variation on a standard growing medium known as "Pfeffer's Solution," and later improved it further to "Knudson C."

This latter formulation has become a commercial product used for orchid seed germination. It contains sucrose, vitamins, and other nutrients and is often supplemented with banana or coconut water. Under sterile conditions, the seeds are sown onto an agar gel containing the Knudson formula in a flask. Interestingly, one of the contaminants Knudson observed that killed the developing embryo was fungi of the genus *Penicillium*, from which the antibiotic penicillin is derived – a fact serendipitously discovered by Alexander Fleming some years later.

In his seminal 1922 article, Knudson makes the following very interesting comment: "In certain experiments Bernard succeeded in germinating seeds of *Cattleya* and *Laelia* without the intervention of the fungus. This was accomplished by using a more concentrated solution of salep." Yes – he is referring to the very salep drink of the preceding chapter, made from powdered orchid tubers. Knudson goes on to say that the tubers contain, "…principally, mucilage 48%, starch 27%, and proteins 5%. It probably contains also some sugar as well as soluble mineral matter. The seedlings obtained in this way were in every respect normal and the germination was very regular." For Knudson, his concluding remark is crucial: "Bernard suggests that some such method might be developed for practical purposes, since the results with the fungus are so unsatisfactory."

We see that Bernard was already considering the idea of asymbiotic germination, and one can almost say that from beyond the grave, Bernard was passing the baton to Knudson and guiding him in his experiments. That the Mediterranean drink salep, which had been consumed for centuries to increase men's libido, played a role in the asymbiotic germination of orchid seeds is truly remarkable (Fig. 2.31).

Fig. 2.31 *In vitro* germination of orchid seeds in various stages of development. Clockwise: protocorms of *Encyclia incumbens*; leaves beginning to sprout on the protocorms of *Encyclia pauciflora*; plantlets of *Cattleya statteriana* (name unresolved) (Images courtesy Thomas Ederer)

At the same time that Bernard was performing his experiments, Professor Hans Burgeff in Germany was independently doing very similar experiments on orchid embryo development. Indeed, he also used salep, but at a very low concentration (2%), and found that *Epidendrum* seeds did not germinate unless appropriate fungi were added. Knudson heavily relied on the contributions of both Bernard and Burgeff but did away entirely with the fungus infection, replacing it with a nutrient-only solution ("solution B") that had, most importantly, added sugars.

In both symbiotic and asymbiotic germination, when the protocorms are ready to be thinned out and replanted, another procedure can be added by dividing them into approximately four segments and regrowing them in a flask. These will again develop into full protocorms and the procedure can be applied again, and again… At some stage, the protocorms can be left to continue growing into rooted plantlets that can subsequently be planted out. Incidentally, the term "protocorm" goes back to the Dutch botanist Melchior Treu and was applied to orchids by Bernard.

Both the above techniques require sterile laboratory conditions not easily achieved by an orchid aficionado. So here we offer an alternative for the amateur, developed by Thomas Ederer of Austria. This technique has been successfully applied to *Epidendrum radicans, Pleione*, and *Disa* seeds. It would certainly be worthwhile experimenting with other genera of orchid seeds in the same fashion.

The first step is to gather some fresh pieces of bark from a forest and cut them up into manageable sizes of say 5 cm in length. Then, put them into the bottom of a clean plastic bottle with the top half cut off. Orchid seeds are sown onto the moistened bark. The top half of the bottle can now be reattached (Fig. 2.32).

The container should now be place in a light area but not in direct sunlight. The bark can be watered daily (rainwater is best), ensuring that they dry out a little between waterings. Within a period of 2–3 weeks, there will be visible protocorms sprouting on the bark! The protocorms sprout leaves and develop into seedlings (Fig. 2.33).

The seedlings should not dry out or be kept wet all the time either. The desirable conditions would be similar to growing small orchids in a terrarium. Once the plantlets have a robust root structure, they can be planted out.

Fig. 2.32 (L): The bark placed in the plastic bottle which can then be sprinkled with orchid seeds. (R): the bottle's top half reattached (Images courtesy Thomas Ederer)

Fig. 2.33 Protocorms of *Epidendrum radicans* sprouting on tree bark (Image courtesy Thomas Ederer)

Fig. 2.34 An orchid seedling (center) on the trunk of a Deodar cedar (Image courtesy Carlos Cruz/flickr/epiphyte78)

Another seed germination method has been successfully employed by Carlos Cruz in California. He starts by soaking *Laelia anceps* seeds in water until they sink. Then, the seeds are squirted with a large syringe onto the rough surface of a tree trunk in close proximity to existing orchid roots. Interestingly, the roots of various other genera seem to work, but of course, the seeds must be in close proximity in order to pick up a suitable mycorrhizal fungus. First protocorms and then small plantlets develop. A similar technique has produced positive results with other genera as well, including *Dendrobium kingianum* [23] (Fig. 2.34).

Of course, these methods are not as effective as asymbiotic germination under laboratory conditions. As mentioned, Bernard's trials were not always successful, and by his own reckoning,

required 50,000 seeds to produce a few hundred plants. In fact, *in vitro* symbiotic germination is now rarely practiced by commercial growers.

The most important artificial technique for the multiplication of a particular orchid is via *meristem* culture, by which a small piece of undifferentiated plant tissue (equivalent to stem cells in humans) is excised from the apex of a new shoot tip and grown *in vitro* on an agar gel culture medium, as for asymbiotic propagation. The resulting plants are all genetically identical copies of the original, and so this technique is employed when clones of a particular cultivar are desired. The plants are also virus free even if the original plant is infected with one, which is an important bonus. The resulting cloned plants of meristem culture are commonly referred to as *mericlones*. Meristem culture is not just for orchid propagation, but also has widespread applications in horticulture and biology.

The history of vegetative cultivation in orchids is rather fraught with challenges and controversy. Initial attempts go back to the end of the 19th century when it was found that nodes from the stems of *Phalaenopsis* could be made to germinate in peat. Modern propagation methods began with the *in vitro* cultivation by Filipino scientist Gavino Rotor at Cornell University in 1949 after listening to a lecture by Knudson. Dr. Rotor used the Knudson C formulation with *Phalaenopsis* stem segments that each contained a node, and which subsequently developed into plantlets.

Other developments in tissue culture were made by the German horticulturalist Hans Thomale together with Dr. Lucie Mayer using various genera, but their work, written in German, remained buried in library archives, as did Rotor's.

It was the French plant scientist Georges Morel who in the early 1960s made the notion of meristem propagation popular with his work on *Cymbidiums*. Many object to the fact that he is "generally given exclusive, but completely undeserved … credit for being the first to culture an orchid explant in vitro." [24] Part of the controversy involves the fact that Morel knew of Thomale's work but only cited it years later, and even then, somewhat begrudgingly. Details regarding the whole unseemly saga can be found in the first chapter of Joseph Arditti's monumental opus on plant propagation and article on the same subject [25].

Will the advent of meristem culture make orchids so common that they lose their appeal? "Will they lose the allure and mystique they have retained for centuries, and even travel the way of the Victorian Aspidistra, once fashionable but now forgotten? Perhaps, but as an orchid enthusiast, with more than 90,000 strange and beautiful species and hybrids to choose from – I doubt it." So says Dr. C.J. Schofield of the London School of Hygiene and Tropical Medicine, and so says the author, adding that the number of species and hybrids has risen considerably to over half a million since Schofield's sentiments were written in 1983 [26].

Fig. 2.35 The flowers of the gigantic orchid *Grammatophyllum speciosum* (Image courtesy Reuben C.J. Lim)

Largest and Smallest Orchids

Normally considered the world's largest orchid is the species *Grammatophyllum speciosum*, found throughout Southeast Asia and New Guinea. It can typically grow up to 3 meters in height, although the *Guinness Book of World Records* has one listed as 7.62 meters. Common names for it are "Tiger Orchid," owing to the yellow spotted flowers; the "Sugar Cane Orchid," owing to its resemblance to sugar cane; or the "Queen of Orchids," owing to its grandeur.

It is normally epiphytic, so that the tree branch it grows on will have to support a clustered mass of several hundred kilograms. The waxy, spotted, sometimes fragrant flowers are giants as well, spanning 12–20 cm (5″–8″) across, and a mature plant can sometimes produce up to 10,000 flowers. The Tiger Orchid planted in 1861 at the Singapore Botanic Gardens, has now reached 5 m in diameter and is still going strong (Fig. 2.35).

A *Grammatophyllum speciosum* reportedly weighing two tons was one of the highlights in the 1851 exhibition at Crystal Palace in London. Another orchid with a giant-sized flower is the aptly named *Rossioglossum grande* (see Chap. 3).

Moving from the sublimely large to the ridiculously small, we have the *Campylocentrum insu-*

Fig. 2.36 A microscopic view of *Campylocentrum insulare*, found in Santa Catarina, southern Brazil. It is the current record holder for the world's smallest orchid flower, measuring just 0.5 mm across (Image courtesy Carlos Eduardo de Siqueira)

lare, found in Southern Brazil in 2015. Discovered by orchid researcher Carlos Eduardo de Siqueira on a branch in the greenhouse of his university's botany department, it was initially mistaken for a fungus. Further microscopic examination proved it to be an orchid (Fig. 2.36).

The former smallest orchid was discovered in 2009 in Ecuador and belonged to the *Platystele* genus. It is a giant compared to the *Campylocentrum*, measuring a whisker over 2 mm in diameter – large enough to see without a microscope. Many members of this Latin America genus are also very

Fig. 2.37 Another miniature orchid flower, *Bulbophyllum pygmaeum*, or *Piripiri*, the "Pygmy Tree Orchid," from New Zealand, measures 2.5 mm across (Image courtesy Eric Scanlen)

small and often have translucent flowers. Just slightly larger still is a *Bulbophyllum* flower from New Zealand (Fig. 2.37).

References

1. M.J.M. Christenhusz and J.W. Byng, The number of known plant species in the world and its annual increase, *Phytotaxa*, 261 (3), 201–217, 2016.
2. Gustavo A. Romero-González and Carlos Gómez, Collecting Orchids**,** *ReVisita, Harvard Review of Latin America*, Fall, 2008; *Illustrated World Compendium of Orchids – List of Taxa:* https://worldplants.webarchiv.kit.edu/orchids/statistics.php
3. The name is now properly: International Code of Nomenclature for algae, fungi, and plants (ICN).
4. M.W. Chase, *et al.*, An updated classification of Orchidaceae, *Botanical Journal of the Linnean Society*, 177(2):151–174, 2015, lists 736 orchid genera; A.M. Pridgeon, *et al.*, *Genera Orchidacearum*, Oxford University Press, 1999–2014, lists 765 orchid genera. The IOSPE: http://www.orchidspecies.com/ lists 856 orchid genera. The plant list: http://www.thep lantlist.org/1.1/browse/A/Orchidaceae/ lists 899 orchid genera.
5. Personal communication.
6. Darwin, *Contrivances*, 2nd ed. p.307.
7. R.B. Singer, *et al.*, Orchid Pollinia or Pollinaria for taxonomic identification, *Selbyana* 29(1), 6–19, 2008.
8. The article: J. He, *et al.*, Responses of Green Leaves and Green Pseudobulbs of CAM Orchid *Cattleya laeliocattleya* Aloha Case to Drought Stress, *J. of Botany,* Vol. 2013, 2013, 9 pp, was a study of a CAM *Laeliocattleya* hybrid, but contains other references of non-CAM epiphytes.
9. As quoted in: K. Silvera, Crassulacean Acid Metabolism and epiphytism linked to adaptive radiations in the Orchidaceae, *Plant Physiology*, 149(4), 1838–1847, 2009.
10. https://www.flickr.com/photos/epiphyte78/galleries/72157649154617841/#photo_10571015796
11. B.G. Bag, *et al.*, Study of Antioxidant Property of the Pseudobulb Extract of *Crepidium acuminatum* (Jeevak) and its use in the Green Synthesis of Gold nanoparticles, *Int. J. Res. Chem. Environ.*, 4 (3), 133–138, 2014.
12. E.A. Tsavkelova, Bacteria Associated with Orchids, Chap. 11 in: *Bacteria in Agrobiology: Plant Growth Responses*, Springer, 221–258, 2011.
13. L. van der Pijl, and C.H. Dodson, *Orchid Flowers: Their Pollination and Evolution,* University of Miami Press, Coral Gables, 1966.
14. C. Siegel, Orchids and Hummingbirds: Sex in the fast lane, *Orchid Digest*, Jan., Feb., Mar., 8–17, 2011.
15. *Contrivances* 2nd ed., p.277/278.
16. The term mycorrhizal simply refers to the symbiotic relationship between the fungus and plant.
17. S.E. Smith and D.J. Read, *Mycorrhizal Symbiosis*, Academic Press, New York, 1997.
18. *Contrivances*, 2nd ed. p.278/80.
19. For details: http://www.orchideenvermehrung.at/english/nodes/in%20soil/index.htm.
20. M. Selosse, *et al.*, Noël Bernard (1874-1911): Orchids to symbiosis in a dozen years, one century ago, *Symbiosis,* 54(2), 2011, 61–68.
21. As reproduced in preceding reference.
22. L. Knudson, Nonsymbiotic germination of orchid seeds. *Bot. Gaz.*,73, 1–25, 1922; L. Knudson, A new nutrient solution for germination of orchid seed. *Amer. Orchid Soc. Bull.*, 15, 214–217.
23. See the website: https://www.flickr.com/photos/63394592@N08/12166496555/in/photolist-aejVoX-yLzAqM-smov2C-jx7txZ-jx8A4R-jx9tEs-gfS2Wr-9Mi6rM/
24. J. Arditti, *Micropropagation of Orchids*, John Wiley & Sons, 2009, Vol. 1, p.24.
25. *Ibid.*, Chapter 1, see also the article: History of Orchid Propagation, *AsPac J. Mol. Biol. Biotechnol.*, 18 (1), 171–174, 2010, which is available online.
26. C.J. Schofield, The allure of the orchids, *New Scientist*, 22/29 Dec., 1983, p.914.

Lord Illingworth told me this morning that there was an orchid there as beautiful as the seven deadly sins.

Oscar Wilde (from *A Woman of No Importance*)

Many orchids appear to imitate other life forms in nature, including, ants, birds, bees, flies, wasps, donkeys, ducks, monkeys, spiders, and even people. The latter we have already seen in the case of *Orchis italica.* Some of this imitation is for a very good evolutionary reason and likely not for our own amusement, but nevertheless, we can draw aesthetic pleasure from them. Australia in particular seems to be blessed with a variety of animal imitators (Fig. 3.1).

Anguloa

Anguloas are the orchid version of tulips with the flower sepals wrapped inward, perhaps huddling from the cool temperatures the flower experiences growing at high elevation (1800-2500 m) on the eastern side of the Andes Mountains. *Anguloas* are mostly terrestrial, are related to the *Lycaste* genus, have highly scented waxy flowers, and are deciduous in the winter. The orchid enthusiast can grow *Anguloa,* bearing in mind that they like it cool and that their leaves can be up to a meter long! (Fig. 3.2)

A rare and critically endangered species member endemic to Colombia is that of *Anguloa cliftonii,* which is only found in an 8 sq. km region currently under threat. Many orchid species are under threat in their native habitat, the usual culprit being the *Homo sapien*-related activity.

Anoectochilus

Most orchids have rather ordinary leaves, although they do come in a diverse range of shapes. For the most part, they are solid green. A fine exception to this rule is the genus *Anoectochilus,* whose members not only have exceptional-looking flowers, but vein-patterned leaves that come in dark green or brown colors. Their appearance is not unlike the nerve plant *Fittonia verschaffeltii,* a common houseplant. The orchid name derives from the Greek *anoektos,* meaning "open," and *cheilos,* meaning "lip," referring to the rather frilly lip of the flowers (Fig. 3.3).

Commonly known as jewel or filigree orchids, these small terrestrial orchids are found in India, Southeast Asia, Melanesia, Hawaii, southern China, and Australia. The flowers have strikingly frilly lips with the dorsal sepal forming a hood with the petals.

It is interesting to note that *Anoectochilus formosanus* and *Anct. koshunensis* have been used by the local people in Taiwan to cure snakebite. But it is *Anct. formosanus*, known as "King Medicine" in Taiwan, that has been used extensively in herbal medicinal treatments for all manner of conditions from diabetes and hypertension to cancer. Indeed, a 2005 laboratory study indicates that an extract of *Anct. formosanus* has the ability to inhibit tumor growth in mice [1]. The authors go on to suggest, "It is worth further analyzing the

Fig. 3.1 *Diuris brumalis* is found in Western Australia and known as the "Donkey Orchid" (Image courtesy Justin Brown)

immunomodulating component purified from *Anct. formosanus*, and evaluating its potential value for the treatment of human cancers." In this regard, *The uses and misuses of orchids in medicine,* is a fine article by experimental medicine professor C.J. Bulpitt, and is readily available online [2].

Bulbophyllum

The genus *Bulbophyllum* is the largest in the orchid world, containing around 2000 different species (estimates vary) that cover a wide range of forms and distribution on the warmer continents. It was first described by French aristocrat botanist Louis Marie Aubert Du Petit-Thouars, who lived in exile after the French Revolution. Restricted to Madagascar, La Réunion, and Mauritius, he made good use of his time describing many species of orchids, including 17 of the genus *Bulbophyllum*. The genus has engulfed other genera, such as *Cirrhopetalum,* as well as several species picked off from the *Dendrobiums,* so that its numbers are

fluid; new members are also being discovered in the wild. Papua New Guinea alone has over 600 species, Borneo has over 200, and even Australia has 26 species and New Zealand two, one of which, *Bulbophyllum pygmaeum,* was mentioned in Chap. 2.

Some species have flowers that are pleasantly fragrant, such as *Bulbophyllum ambrosia* and *Bulbophyllum odoratissimum*, while others are highly malodorous (see Chap. 5).

In *Bulbophyllum*, the leaf comes directly out of the pseudobulb, thus giving the genus its name, derived from the Latin *bulbus* and the Greek *phyllon,* meaning leaf. The flowers of several species have very elongated lateral sepals that are joined for some of their length, sometimes in clusters (Fig. 3.4).

One recent addition to the genus is *Bulbophyllum nocturnum*, discovered in 2011 in New Britain, an island of Papua New Guinea [3]. It flowers for only a single night, with opening hours from approximately 10 pm to 10 am. A very unusual feature of this orchid's flower are the dangling appendages that have a resemblance to the fruiting bodies of slime molds. These could play some role in attracting its as yet unknown nocturnal insect pollinator (Fig. 3.5).

Of course, not all *Bulbophyllums* are so bizarre in appearance, so included is a picture of the little Himalayan *Bulb. guttulatum* for contrast (Fig. 3.6).

Cattleya

Cattleyas, mostly found in South America, come in many colors, from lavender to pink to white with various colors on the lip. *Cattleya* were once distinguished from another genus *Laelia* by their varying number of pollinia in their flowers: *Laelia* had eight pollinia, while *Cattleya* had four. Yet, this distinction no longer holds: due to recent DNA evidence, the large-flowered Brazilian *Laelia* has been reclassified as *Cattleya*. Thus, this latter genus has risen in number, and it now contains species with both 4 and 8 pollinia. Another genus, *Sophronitis*, first described by Lindley in 1828 and also having 8 pollinia, has likewise been

Fig. 3.2 (L): The rather unusual flowers of *Anguloa uniflora*. (R): The rare and endangered *Anguloa cliftonii* (Images courtesy (L) nolehace photography, (R) Kevin Holcomb)

Fig. 3.3 (L): Typical of the genus is the frilled lip of *Anoectochilus albolineatus*. (R): The patterned leaves of *Anoectochilus monicae* (Images courtesy (L) Nova de Jong, (R) John Varigos)

absorbed into *Cattleya,* swelling its numbers to over 180 species.

The *Cattleya* genus is one of the quintessential orchid types in the mind of the public, taking on fame as the "corsage orchid" of post-war decades. Its flowers' striking features, together with an intoxicating perfume, are always a crowd pleaser at any orchid show. The genus is very popular in

Fig. 3.4 Clockwise from top right: A cluster of *Bulbophyllum odoratissimum* flowers, which have a pleasant fragrance; *Bulb. longistelidium; Bulb. dayanum; Bulb. hirundinis* (Images courtesy (clockwise): C. Solimbero for www.Hortus Orchis.com, Joost Riksen, Piotr Markiewicz, Steve Beckendorf)

the production of hybrids and is crossed with many different genera (Fig. 3.7).

Unlike nearly all *Cattleya* orchids, whose inflorescence comes out from the top of the pseudobulb, there are two species, *C. nobilior* and *C. walkeriana*, where the flower stem emanates from the base of the pseudobulb. The flowers of both are similar in appearance and fragrant but differ in chromosome count. Both are very popular in their native Brazil as well as in Japan, which has orchid societies devoted solely to their cultivation.

Coelogyne

A new species of *Coelogyne* was found on the southern Philippine island of Mindanao in 1999 and first described in 2001. The epiphytic *Coelogyne usitana* was named after its orchid collector finder Villamor T. Usita, who discovered it growing on tree branches and rocks. The whitish petals and sepals contrast with the lip color, which can vary from brown or chocolate shades to deep orange or red. As is typical with many *Coelogynes*, the flower tilts downward. Although each spike only has one flower open at a time, the blooms open sequentially for up to a year or more (Fig. 3.8).

Coelogyne is a very old genus established by Dr. John Lindley in 1821. Many of the approximately 200 species are cool-growing from Himalayan and mountainous Southeast Asian regions, in contrast to *Coel. usitana,* which is warm to hot-growing. The name derives from the Green words *koilos,* meaning "hollow," and *gyne,* meaning "female," referring to the deep cavity that houses the stigma (Fig. 3.9).

One of the most floriferous of any species is *Coelogyne cristata,* which not only has delicately beautiful flowers with deep yellow-throated lips but

Fig. 3.5 The flower of the single night-blooming *Bulbophyllum nocturnum* with its unusual appendages hanging down (Image courtesy Jaap Jan Vermeulen, from the article cited in the text [3])

Fig. 3.6 *Bulbophyllum guttulatum,* found in China and the Himalayan regions (Image courtesy Norbert Dank/ www.flickr.com/photos/nurelias)

also grows to an enormous size that can produce a veritable snowstorm of such blossoms (Fig. 3.10).

One interesting aspect to some *Coelogyne* species is their long pendulous inflorescence supporting large numbers of flowers, as is the case for *Coelogyne massangeana.* Such a beautiful name was given to this beautiful species by Reichenbach in 1878, but now the less exciting name *Coelogyne tomentosa* is preferred. The long cascading clusters of flowers have an intoxicating fragrance that transports one to the montane forest regions of Malaysia, Sumatra, and Borneo. It was the first orchid grown by the author and remains a favorite. Unfortunately, due to the long pendulous flower stems that can have up to 30 blooms (not particularly long lasting), they are difficult to transport to flower shows for display. Even so, they are well worth growing, as they require only cool, shaded conditions and a bit of nutrients every now and then in order to thrive.

A similar sounding genus is *Coeloglossum,* consisting mainly of a widespread terrestrial species *Coeloglossum viride,* now classified as *Dactylorhiza viridis.* It is completely unrelated to the *Coelogyne* genus, and although rather unprepossessing in appearance, an extract from the plant has shown some promise ameliorating brain tissue damage in rats. One compound found in the extract, *Dactylorhin B,* has been shown to reduce the toxic effects on rat neurons of beta-amyloid fragments, which are associated with Alzheimer's disease [4] (Fig. 3.11).

Cymbidium

The genus *Cymbidium* takes its name from the Greek word *kymbós,* referring to the boat shape of its lip. It has been known in China since before 500 B.C. and is found in various Asian countries as well as northern Australia. The waxy flowers of modern hybrids can last more than 2 months with over a dozen flowers on a single stem, making the plants highly prized in the cut-flower market. The genus was first described in 1799 by Olof Swartz [5].

The genus's growth habit can be epiphytic, terrestrial, or lithophytic; *Cymbidium floribundum* can be found in all three habitats. The number of species in this genus is still unresolved, with more than 80 species currently accepted. Many from

Fig. 3.7 The alluring beauty of *Cattleyas*. Clockwise: *Cattleya tenebrosa; C. warneri; C. porphyroglossa; C. schroederae;* and the *C. nobilior*, which has its inflorescence, like *C. walke-riana*, emanating from the base of the pseudobulb, unlike other *Cattleya* orchids (Images courtesy Christian Furtwängler, and Antonio Donatini Neto (*C. nobilior*))

Fig. 3.8 The recently discovered Philippine orchid *Coelogyne usitana* (Image courtesy Peter Tremain)

China, such as *Cym. ensifolium* and *Cym. sinense,* are notable for their fragrance. In India, all parts of the *Cymbidium* – rhizomes, leaves, flowers, pseudobulbs – have been used as herbal medications since the earliest times (Fig. 3.12).

Some *Cymbidium* orchids display a compact form enjoyed for their delicate beauty, such as *Cym. floribundum*, the "Spring Orchid," *Cym. goeringii*, and *Cym. tigrinum*, the latter two also being fragrant. While the flowers of *Cym. floribundum* are not scented to humans, they emit two hydroxyl fatty acids that mimic bee secretions and attract the lone oriental honeybee (*Apis cerana*) or even entire swarms [6]. This discovery could prove to be useful in beekeeping. Another scented *Cymbidium* is *Cym. mastersii*, which has high commercial value but is becoming increasingly vulnerable to loss of habitat and human predation.

Fig. 3.9 Clockwise: *Coelogyne xyrekes, Coel. celebensis, Coel. sparsa, Coel. tenompokensis* (Images courtesy Steve Beckendorf *(Coel. xyrekes)*, all others Piotr Markiewicz)

Fig. 3.10 Two very different growth habits of *Coelogyne*. (L): A very mature *Coel. cristata* with its massive display of gorgeous white flowers. (R): The intoxicatingly fragrant, cascading flowers of *Coel. massangeana* (Images courtesy (L) Steve Beckendorf, (R) Andrew Orme)

Fig. 3.11 *Coeloglossum viride* contains important chemicals that help improve the function of damaged brain tissue (Image courtesy José Luis Santamaría)

Cypripedium

Cypripedium, derived from *cypros*, meaning "Cyprus," and *pedilon*, meaning "slipper," is composed of perennial terrestrials with a creeping rhizome, forming clumps of plants. Its 58 species grow in temperate woodlands throughout Asia, Europe and the Americas. Many hybrids are easily grown as garden plants, as they tolerate most soil conditions, and some species can even tolerate snow in the winter, such as *Cypripedium guttatum,* which grows in Alaska and Siberia and other places of extreme cold. In fact, such conditions are necessary for its growing success, and the reader should not be tempted to grow them unless living in a sufficiently cold growing zone. Of course, such growing conditions are highly unusual for any orchid species (Fig. 3.13).

Fig. 3.12 Clockwise from left: *Cymbidium mastersii; Cym. insigne; Cym. tracyanum; Cym. erythrostylum* (Images courtesy Ron Parsons (*Cym. mastersii*), Sylvia Kappl all others)

Fig. 3.13 (L): The delicate beauty of *Cypripedium reginae. (*R): *Cypripedium calceolus* (Image courtesy (L) David McAdoo, (R) José Luis Santamaría)

Dendrobium

Dendrobiums form one of the largest orchid genera, with about 1500 species spread throughout Asia and the South Pacific. The name derives from the Greek *dendron*, meaning tree, and *bios,* meaning life, referring to the epiphytic growth habit of many species members. This is another genus described by Swartz in his *Nova Acta Scientiarum Regiae Societatis Upsaliensis* (1799).

Many *Dendrobium* species have been used in traditional Chinese medicine for centuries, with the dried stems made into tonic or given as treatment for various ailments including cancer. Possibly stemming from the mythical teachings of Shen Nung mentioned in Chap. 1, the Chinese pharmacopoeia from 200 BC, *Shen-nung Pen-tsao Ching,* describes *Dendrobiums* as a tonic, astringent, analgesic, and anti-inflammatory. So, it is only natural that modern medical research would look into this genus, and the findings indicate that it has many pharmacological properties, including antioxidant, anti-inflammatory, anti-tumor, reversing multi-drug resistance, and antimicrobial/fungal effects. The compound Denbinobin, found in the beautiful

Dendrobium nobile, has been found to impair the spread of prostate cancer (2014); the compound erianin from *Den. chrysotoxum* is lethal to human leukemia cells (2013); there are antitumor effects from polysaccharides in *Den. candidum and Den. nobile* (2010); and the compound moscatilin, extracted from the stems of *Den. loddigesii,* inhibits lung cancer motility and invasion (2013), induces the death of human esophageal (2013) and colorectal cancer cells (2008), and inhibits the spread and metastasis of human breast cancer (2013). [Dates in parenthesis indicate the year of the research study as cited in, *Cancer Inhibitors from Chinese Natural Medicines* [7]] (Fig. 3.14).

Dendrobium flowers can be used in cooking. In Thailand, they are fried with butter and eaten, while in the Chinese province of Yunnan, *Dendrobium chrysotoxum* flowers are made into tea. Researchers have also investigated this orchid's ability to prevent retinal inflammation from diabetes [8]. *Den. chrysotoxum* is interesting, as the plain yellow flowers exhibit a dramatic bullseye in the near ultraviolet, thought to be an aid in pollination. Some other *Dendrobium* species also have this feature [9].

Fig. 3.14 Within the canes of *Dendrobium loddigesii* is the compound moscatilin, shown to possess powerful anticancer effects (Image courtesy Wiel Driessen)

Various *Dendrobium* species have normal-looking ovate pseudobulbs, while others have developed into long cane structures. It is natural with such a large genus that there will be some rare and exotic gems.

Dendrobium antennatum is unusual in that its greenish petals are swept upward and twisted, inspiring its common names "Antelope Orchid" or "Antenna Orchid." The dorsal sepal also has a distinctive curl to it. Found in Queensland, Australia, Papua, and the Solomon Islands, it has cane-like pseudobulbs and fragrant, long-lasting flowers in summer that can grow up to 7.5 cm in length. It is listed as an endangered species in its native habitat, but fortunately, it is popular with orchid growers and does well.

The Australian *Dendrobium teretifolium*, also classified as *Dockrillia teretifolia,* has long, pencil-thin leaves up to 60 cm (24″) in length. The pendulous growth habit can be up to 3 m in length, with clusters of fragrant, white, spidery flowers that make for an extraordinary display.

One of the most bizarre orchid flowers of all belongs to *Dendrobium spectabile.* Found in Papua

and New Guinea, the Solomon Islands, Vanuatu, and Australia, the petals and sepals are contorted and twisted with burgundy colored stripes and spots, are 8 cm across and have an intense fragrance. The back of the labellum is hinged, with the canes growing up to a meter (3 ft) in length. A mature plant covered in these blooms is a stunning sight to behold.

Dendrobium unicum, or "Unique Dendrobium," is a small species endemic in Southeast Asia, with flowers twisted in different orientations up to 10 cm (4 in) that have curled back petals and sepals. The scent of the fragrant blooms has been described as that of "tangerines and sharpened color crayons" if that is possible to imagine. This species was described only recently in 1970 by Seidenfaden (Fig. 3.16).

Two other *Dendrobium* species that have elongated, twisted petals are *Den. stratioides* and *Den. lasianthera.* The latter is found in the hot and humid forests of New Guinea growing as an epiphyte with canes up to 2–3 m (6-10 ft) in length. The flowers are fragrant, long lasting (2 months), and have somewhat variable colors. Given these circumstances, one cannot easily grow them in captivity. But what a dazzling sight! (Fig. 3.16)

The *Dendrobium* genus nearly lost a member when the genus of *Winika* was created for its sole member, *Winika cunninghamii,* formerly classified as *Den. cunninghamii* [10]. The name *Winika* is the old Maori name for this orchid, which is endemic to New Zealand and commonly found in rainforests. The name derives from the sacred canoe *Te Winika* of the Tainui people, whose hull was formed from a tree upon which this orchid was growing.

The epiphytic orchid has thin, cane-like stems from which the leaves and flowers grow (Fig. 3.17).

However, the new genus has not been widely accepted, and the species in standard references is listed by its former *Dendrobium* classification, with *Winika cunninghamii* mentioned as a synonym.

Dendrophylax

There is a specter lurking on trees in Florida and parts of the Caribbean, namely the endangered epiphyte *Dendrophylax lindenii* (syn. *Polyrrhiza lindenii*), or the "Ghost Orchid." It is named after

Fig. 3.15 Clockwise: *Dendrobium densiflorum; Den. tobaense, Den. polysema; Den. stratiotes* (Images courtesy C. Solimbero for www.Hortus Orchis.com)

its discoverer, the famed Belgian botanist and horticulturist Jean Jules Linden, who originally found it in Cuba.

While the Ghost Orchid is difficult to cultivate outside its natural environment, this little stunner has been the subject of intense interest – not always favorable – as seen in Susan Orlean's *The Orchid Thief*. Because of its precarious status, it must be studied *in situ* [11] (Fig. 3.18).

Atypically, the orchid plant has no leaves and instead consists of a mass of roots (hence the synonym name, *Polyrrhiza* – "many rooted" in Greek) affixed to a tree trunk, which also perform the task

of photosynthesis. The flowers have a "sweet smelling and somewhat fruity" fragrance with a long nectar-containing spur. As in such cases, the pollinator should be a large moth, and here, it is the giant sphinx moth *Cocytius antaeus*, the only insect in North America with a sufficiently long proboscis. But this moth is also rare in the orchid's habitat, resulting in the low prevalence of this species.

There are 14 species of this genus, all found in the Caribbean region, and they are distantly related to the African genus *Angraecum*, which they resemble.

Fig. 3.16 Top: It is all about the lip, which presents in various orientations in the unusual *Dendrobium unicum.* Bottom*:* The dazzling flowers of *Dendrobium lasianthera,* a hot-growing species from New Guinea (Images courtesy (top) Joost Riksen, (bottom) Eric Hunt)

Dracula

One of the most bizarre of all orchid genera is *Dracula,* also known as "Vampire Orchid" or "Monkey Orchid." The very name itself conjures up weird images. The plant is distributed widely in cloud forests throughout Mexico, Central, and South America and consist of over 120 species. Most species are epiphytic, although they lack pseudobulbs and the flowers are generally small in size. Previously classified as *Masdevallia,* the separate genus *Dracula* (meaning "little dragon" owing to their bat-like appearance) was created in 1978 but was *not,* contrary to expectations, named after everyone's favorite bloodsucking Count. The three sepals have long, thin tails, the lip rounded

Fig. 3.17 The very delicate flower of *Dendrobium* (*Winika*) *cunninghamii,* endemic to New Zealand (Image courtesy Michael Pratt)

Fig. 3.18 The "Ghost Orchid," *Dendrophylax lindenii,* found in Cuba and Southwest Florida. Note the long nectar spur arching behind the flower and the large root mass (Image courtesy David McAdoo)

Fig. 3.19 The bizarre-looking members of the genus *Dracula*, clockwise from top: *Drac. diana; Drac. lafleurii* 'John Leathers'; *Drac. levii; Drac. vampira* (close up) (Images courtesy Ron Parsons (bottom row), Luis Carlos Maya, Sociedad Colombiana de Orquideología (top))

and protruding, and the petals are small, often dark and wart-like, which results in the flower having a distinctly simian appearance. These orchids should not be confused with another orchid genus with simian features, *Orchis simia* (Fig. 3.19).

The mushroom-like labellum serves a purpose, in that it also smells like a mushroom to attract fruit flies, and in the courting and mating activity that subsequently takes place in the flower, pollination occurs [12].

Epidendrum

Like *Dendrobium*, the genus *Epidendrum* has more than 1500 different species. Historically, *Epidendrum* contained species that are now classified as other genera, as it was a catchall for any epiphytic orchid. The name means "upon trees," referring to their growth habit, but some are not actual epiphytes, like *Epi. radicans,* which grows along roadsides and in the author's garden. They

Fig. 3.20 The non-resupinate alpine *Epidendrum fimbriatum* can be found at elevations of 3200 m (10,000 ft) in parts of South America (Image courtesy Jorge J. Restrepo)

are found in South, Central, and lower regions of North America in a variety of habitats, including alpine climates, as the *Epidendrum fimbriatum* is found growing in high mountainous parts of South America (Fig. 3.20).

Many *Epidendrums* have the labellum attached to the sides of a protruding column, taking many unusual shapes. One such is the rather bizarre *Epidendrum ilense,* which is severely endangered where it was discovered in Ecuador in 1976 but which can now be purchased commercially, owing to artificial propagation. An interesting feature of *Epidendrum radicans* is its variable chromosome numbers, ranging from 2n = 40 to 2n = 64. One specimen of *Epi. ibaguense,* which is very similar in appearance to *Epi. radicans,* was found to have 2n = 70, whereas humans have 2n = 46 chromosomes [13]. Both *Epi. radicans, Epi. ibaguense,* and *Epi. secundum,* are considered "crucifix orchids" owing to their flowers' form. All three can be found in the colors, red, orange, yellow, and lavender, with *Epi. secundum* having the largest of all orchid seeds, measuring 6 mm (nearly one-quarter inch) (Fig. 3.21).

One *Epidendrum* with a most curious-sounding name is *Epidendrum pseudepidendrum,* an epiphyte from Costa Rica and Panama. In 1852, Reichenbach originally classified it as *Pseudepidendrum spectabile,* perhaps thinking it was not quite a true *Epidendrum,* but he reclassified it under its current name in 1856. One would have thought that a more preferable change of name would have been to *Epidendrum spectabile,* but alas, that name was already taken in 1853 by Focke for an orchid that now bears the name *Encyclia granitica* (Fig. 3.22).

Epipogium

Besides the swampy forests of Florida, Cuba, and the Caribbean, where lurks the ghostly presence of *Dendrophylax lindenii,* another Ghost Orchid can sometimes be seen in various parts of Europe, namely *Epipogium aphyllum.* It differs from its epiphytic New World counterpart in that it is terrestrial, but it is similar in that it too is rare and endangered in its native habitat. This is due in part to its unusual growth habit, whereby the plant does not create food from photosynthesis, but rather from a parasitic relationship (*myco-heterotrophy*) with various fungi that are associated with the roots of particular conifer trees. The leafless stems actually grow underground and only appear above the surface when in flower. Thought to be extinct in Britain, one plant appeared in 2009, and they rarely exist at a site for more than 10 years.

The genus *Epipogium* has but three species scattered through Europe, Asia, Africa, and parts of the Pacific (Fig. 3.23).

Habenaria

The genus *Habenaria* takes its name from the Latin *habena,* meaning "strap" or "rein," which refers to the long thin petals and lip of many of its species. In spite of there being more than 600 mainly terrestrial species found on all continents but Antarctica, they are rare in cultivation. Their most notable member is the delicately beautiful *Habenaria radiata* (*Pecteilis radiata*), or "Flying

Fig. 3.21 Clockwise from top left: *Epidendrum ibaguense,* shown here in its lavender form, it can also be found as red, yellow, or orange. It is very similar in appearance to *Epi. radicans,* which also comes in the same four colors; *Epi. ilense; Epi. ciliare;* and *Epi. stamfordianum* (Images courtesy (clockwise) Sylvia Kappl, Norbert Dank, Mabelín Santos, Norbert Dank/ www.flickr.com/photos/nurelias)

Egret Flower" from Japan, Korea, and parts of China and Russia (Fig. 3.24).

Visible are the two pollinia on each side of the column with a hole leading to a curved nectar-containing spur, which can just be seen in the figure. Most *Habenaria* are pollinated by moths or butterflies with an appropriately long proboscis to reach the nectar reward. In the case of *Hab. radiata,* the culprit is the brown skipper butterfly *Parnara guttata.*

This latter orchid species is especially appreciated in Japan, where it is named *sagisou* (Heron Plant). Ironically, its habitat is also home to the white egret bird. Unfortunately, this orchid is becoming increasingly rare in the wild, as its habitat is under threat from development.

Hexalectris

While researching a book on the wild orchids of Arizona and New Mexico, Dr. Ronald Coleman came across an orchid in Arizona that was originally thought to be a member of the species

Fig. 3.22 The unusually named *Epidendrum pseudepidendrum* from Costa Rica and Panama (Image courtesy Norbert Dank//www.flickr.com/photos/nurelias)

Fig. 3.23 The very rare and elusive "Ghost Orchid," *Epipogium aphyllum* (Image courtesy Hans Stieglitz/Wiki Commons)

Hexalectris spicata. Later, he concluded that it was actually *Hexalectris revoluta,* which it closely resembled. Subsequent DNA analysis determined that it was in fact a new species altogether, and it has since been named *Hexalectris*

colemanii [14]. It is one of the ten known members of this terrestrial genus and happens to grow in a few canyon regions of Arizona, U.S.A. It also goes by the name "Coleman's Coralroot." It is a very rare species under threat from the human species, and it is another orchid that obtains its nutrients from a parasitic relationship with a specific mycorrhizal soil fungus, as the plant has no leaves and lacks chlorophyll for photosynthesis (Fig. 3.25).

Lepanthes

Lepanthes are found in cloud forests of Mexico through Central and South America. The name *Lepanthes* is derived from the Greek, *lepis* meaning "scale," and *anthos* for "flower," owing to their scale-like appearance. The rather small flowers exhibit two interesting traits, although not in all cases. In the first instance, many of the flowers lie prostrate right on the leaf surface with the leaf sometimes patterned, as is the case for *Lepanthes saltatrix* or *Lths. gargoyla* (Fig. 3.26).

The second unusual habit is that flowers can be densely packed as if strung out like cloths on a washing line, as in the case of *Lepanthes acuminata* (Fig. 3.27).

Of the more than 800 species of *Lepanthes,* there are more conventional flowering types, such as the Bolivian *Lepanthes nycteri* and *Lepanthes caprimulgus* from Ecuador and Peru, both of which dangle on thin stems away from the leaf (Fig. 3.28).

How this non-rewarding orchid (in the pollinator sense) gets itself pollinated is another story, discussed in Chap. 4. With their vast array of forms and growth, one could spend an entire lifetime studying this unusual genus.

Masdevallia

The genus *Masdevallia* grows in the same cool New World tropical regions as *Dracula,* and it was named for the eighteenth century Spanish doctor and botanist, José Masdevall. The good doctor

Fig. 3.24 (L) The Flying Egret Orchid, *Habenaria radiata.* Note the hole leading to the nectar spur, just noticeable at the back of the flower. (R) *Habenaria rhodocheila* (syn. *Hab. roebelenii*), from parts of Southeast Asia and southern China, comes in a variety of soft reds, oranges, yellows, and pinks. It has a curved nectar spur similar to the one found on *Hab. radiata* (Images courtesy (L) Tom Velardi), (R) Nova de Jong)

Fig. 3.25 (L): The rare *Hexalectris colemanii,* found in only a few locations in Arizona, U.S.A., and (R): the more geographically widespread *Hexalectris spicata* (Images courtesy (L) Ronald Coleman, (R) Alan Cressler)

Fig. 3.26 *Lepanthes saltatrix* (L) and *Lths. gargoyla* (R) produce flowers that lie on the leaf surface (Images courtesy Wiel Driessen)

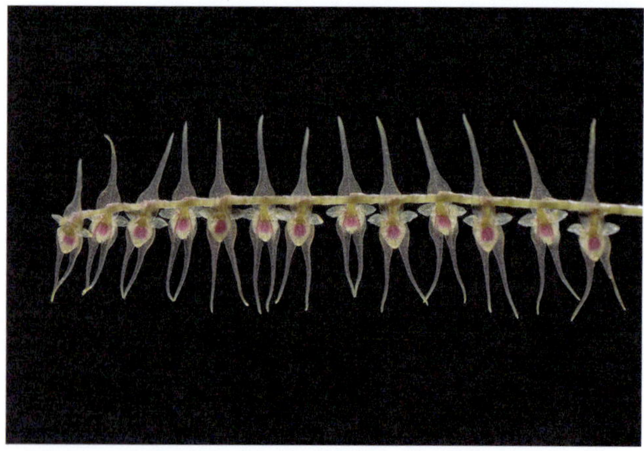

Fig. 3.27 *Lepanthes acuminata* has a row of flowers densely packed along the inflorescence (Image courtesy C. Solimbergo for www.HortusOrchis.com)

became famous for devising his *Antipyretic Opiate*, which, coupled with other medicinal measures, proved to be effective in combating a Spanish epidemic of "putrid and malignant fevers."

Comprising more than 500 species, these charming little flowers have joined sepals at least for part of their length, with long, thin tails, and very small petals that come in a wide variety of vivid colors. Their growth can be either epiphytic, lithophytic, or terrestrial, and like *Draculas*, they have no pseudobulbs. Most are found at rather high altitudes, and some, like the epiphytes *Masdevallia rosea* and *Masd. schlimii,* can grow at altitudes over 10,000 ft. (3200 m).

Owing to their cloud forest natural habitat, they need to be kept moist at all times, and many species can be difficult to maintain in cultivation. Provide insufficient moisture, humidity, temperature, or too much light, and the plants will drop leaves or turn yellow. The author knows this from experience. These are not really an orchid for the novice, but if grown properly, they are certainly very endearing to look at. Indeed, at the height of Orchid Mania, this exotic genus was highly sought after, and plants of *Masdevallia tovarensis* with its virginal white flowers were selling for one guinea per leaf – a small fortune at the time (Fig. 3.29).

Fig. 3.28 (L): *Lepanthes nycteris* and (R): *Lths. caprimulgus* appear on long, thin stems from the leaf base (Images courtesy Joost riksen)

Fig. 3.29 Clockwise top left: *Masdevallia caloptera; Masd. pinocchio* (center)*; Masd. hirtzii; Masd. harlequina; Masd. tovarensis,* once costing a small fortune*; Masd. oxapampaensis* (Images courtesy (clockwise): Ron Parsons, Eric Hunt (center), Wiel Driessen, Ron Parsons, Van Swearingen, Steve Beckendorf)

Fig. 3.30 The epiphytic *Dryadella zebrina*, whose leaves are tinged with purple, exhibits the typical clustered Dryadella growth habit (Image courtesy Sylvia Kappl)

Dryadella, formerly classified as *Masdevallia* until the genus was split off by Carlyle Luer in 1978, consists of about 50 species from Latin America. They form clumps with flowers, often bearing purple or maroon spots and appearing singly on short stems. A *dryad* in Greek mythology was a "tree nymph," and *ella* the diminutive, so this is a perfect description of these little tree-dwelling orchids (Fig. 3.30).

Maxillaria

This large genus of approximately 320 mostly epiphytic species grows in the tropical and subtropical Americas, with one species, *Maxillaria parviflora* (*Camaridium vestitum*), also found in Florida, U.S.A. The name derives from the Latin *maxilla*, meaning "jaw," as the lip and column resemble an insect jaw -- at least according to Spanish botanists Hipólito Ruiz López and José Antonio Pavón Jiménez, who first described the genus in 1794.

Each inflorescence produces only a single flower, but some species compensate by producing numerous inflorescences at the base of a pseudobulb. The diversity of forms and colors is remarkable (Fig. 3.31).

In 2007, some 46 species were hived off to create the new genus of *Maxillariella*, the diminutive

"iella" owing to their often smaller flowers [15], although this classification has not been unanimously accepted. Also in the same year, the new genus *Brasiliorchis* was created, of which *Maxillaria picta* is the type specimen [16]. Numerous other *Maxillaria* were similarly reclassified as *Brasiliorchis*. This new classification has not been universally accepted either: on the Kew World Checklist, *Brasiliorchis* is not an accepted genus name, but on The Plant List, it is. Thus, you will find *Maxillaria picta* as the accepted name on the Kew World Checklist, with *Brasiliorchis picta* and *Bolbidium pictum* as synonyms. On The Plant List, *Brasiliorchis picta* is the accepted name, with *Maxillaria picta* and *Bolbidium pictum* as synonyms. The only thing these two lists have in common is that *Bolbidium pictum* is not an accepted name on either one. The *Bolbidium pictum* taxonomic aspect will not be discussed here. It's complicated [17].

In any event, the brownish-purple-spotted yellow flower of this orchid is very beautiful, and even the leaves are an attractive pale green (Fig. 3.32).

Miltonia/Miltoniopsis

Confusion abounds regarding these two genera, and for good reason. The *Miltonia* genus was created by Dr. John Lindley in 1837 based on the type species *Miltonia spectabilis,* found in eastern Brazil, although other species members are found in Argentina and Paraguay. Lindley named it after Viscount Milton (real name: Charles William Wentworth Fitzwilliam), a British aristocrat who was the inaugural President of the British Association for the Advancement of Science (1831), a politician, and an orchid lover. Species of the genus now denoted as *Miltoniopsis* were also classified as *Miltonia*.

Currently, there are 20 accepted *Miltonia* species, eight of which are natural hybrids. Besides the clean-out of the *Miltoniopsis*, other *Miltonia* species were transferred in 1983 to a new genus *Miltonioides* but have ended up in the *Oncidium* genus, with *Miltonioides* a synonym (Fig. 3.33).

Fig. 3.31 Clockwise: *Maxillaria neglecta; Max. striata; Max. (Brasiliorchis) chrysantha; Max. coccinea; Max. (Brasiliorchis) picta* (Images courtesy, clockwise: Wiel Driessen, Guido Deburghgraeve, Wiel Driessen, Wiel Driessen, Ruud de Block)

Fig. 3.32 *Maxillariella vulcanica.* It is similar in appearance to *Maxillariella arbuscula*, but the former has spotted pseudobulbs whereas the latter does not (Image courtesy Naoki Takebayashi)

We now move on to *Miltoniopsis*, whose name was proposed long ago by the French botanist Alexandre Godefroy-Lebeuf in the 1889 issue of his own journal, *L'Orchidophile* [18]. Its species members were *Miltoniopsis vexillaria, Mltnps. roezlii, Mltnps. phalaenopsis*, and *Mltnps. warszewiczii*. He got it spot on, but what was glaringly obvious to Godefroy-Lebeuf simply did not take root in the orchid world, with the name only appearing sporadically over the years. The main classification of *Miltonia* persisted until 1976, when Dunsterville and Geray firmly re-established the long neglected genus of *Miltoniopsis* and adding another genus member, *Miltoniopsis santanaei* [19]. There has been one further species addition, *Miltoniopsis bismarckii*, found in 1985 by collector Klaus von Bismarck growing at 1000 m in the Peruvian rainforest.

Miltoniopsis species are found in Central and western parts of South America, and it is these, not the *Miltonias*, that are known as the "pansy orchid (Fig. 3.34)."

So what are the differences between a *Miltonia* and a *Miltoniopsis*? There is certainly a morphological difference in the petals and sepals, in that *Miltoniopsis* orchids do indeed look like pansies

Fig. 3.33 Clockwise: *Miltonia spectabilis* (the type speci-
men)*; Milt. clowesii; Milt. moreliana; Milt. kayasimae*
(Images courtesy Kath Knight (*Milt. spectabilis*), (next two)
Norbert Dank/www.flickr.com/photos/nurelias/, Ian Fraser
(*Milt. kayasimae*))

and *Miltonia* orchids do not. *Miltonia* plants have
two leaves coming out of each pseudobulb of a
yellow-green hue, whereas *Miltoniopsis* plants
have only one leaf of a blue-green hue. The former is warmer growing (referred to as "Brazilian
type"), the latter cooler ("Colombian type").
Further, *Miltonia* pseudobulbs are spaced apart
along a creeping rhizome, unlike *Miltoniopsis*
pseudobulbs, which form in clusters. More
nuanced differences exist as well.

Fig. 3.34 Clockwise: All six members of the genus *Miltoniopsis* (clockwise top left): *Mps. bismarckii; Mps. phalaenopsis; Mps. roezlii; Mps. santanaei; Mps. vexillaria; Mps. warscewiczii* (Images courtesy (first four), Ron Parsons; Nicolás Gómez (*Mps. vexillaria*) and Gustavo Adolfo Aguirre (*Mps. warscewiczii*), Sociedad Colombiana de Orquideología)

Over the years, thousands of stunning crosses have been made with both genera, starting with *Miltoniopsis* Bleuana (*vexillaria × roezlii*), made by the Frenchman Alfred Bleu in 1889 (Fig. 3.35).

Odontoglossum

It was George Ure Skinner, whom we have met earlier, who first succeeded in sending *Odontoglossum bictoniense* (now known as *Rhynchostele bictoniense*) from Guatemala to England. It was subsequently named for the vil-

lage of Bicton in Devon, England, where it was first cultivated and bloomed in 1836 (Fig. 3.36).

Another discovery of Ure Skinner was *Odontoglossum grande* (now *Rossioglossum grande*), whose very large flowers can reach the size of a dinner plate! The unusual name of the genus derives from the Greek *odontos*, meaning "tooth," and *glossa,* meaning "tongue," referring to the tooth and tongue-like appearance of the lip. Many members of this genus have been transferred out to form other genera, such as *Otoglossum* or *Rossioglossum*, or simply reclassified with existing genera, such as *Oncidium* or *Miltonia*.

Fig. 3.35 The modern hybrid *Miltonia* Hajime Ono × Robert Jackson (Courtesy Christina Boris/orchidfetish.com/)

Fig. 3.36 The very large flower of the epiphytic *Rossioglossum grande* ("Tiger Orchid"), found in Central America and southern Mexico (Image courtesy Christian Furtwängler)

Thus, according to some taxonomists, this genus no longer exists, yet it has historically served a very useful function, and we will continue with this tradition. The species members are striking in appearance with many showing a multi-flowered inflorescence with colored blotches on white or yellow backgrounds. It is easy to assume that these colored patterns must be the result of hybridization, but they are simply the handiwork of Mother Nature. In addition, there are many natural hybrids, such as *Odontoglossum × andersonianum*, which is a natural cross between *Odm. crispum* and *Odm. gloriosum*.

Odontoglossums have also been interbred by humans with several other genera, such as *Brassia (Odontobrassia), Cochloida (Odontioda), Miltonia (Odontonia),* and *Oncidium (Odontocidium)* (Fig. 3.37).

Sometimes, you do not have to go into a remote jungle to discover an orchid new to science. Belgian medical doctor, orchid enthusiast, and *Odontoglossum* specialist Dr. Guido Deburghgraeve found a new species while visiting the Ecuadorian orchid establishment Ecuagenera. "They were labelled as *Odontoglossum helgae,* but of course they were not." The new species was subsequently named by Dalström and Merino in 2010 as *Odontoglossum deburghgraeveanum*. Although similar to *Odontoglossum harryanum, Odm. helgae,* and *Odm. wyattianum,* there are various anatomical differences [20]. The new species has been found growing in the Zumba area in Southern Ecuador as well as near Macho Picchu. It is certainly an orchid to look out for if visiting this historic site.

Oncidium

The *Oncidium* story goes back to Dutch-born scientist Nikolaus Joseph von Jacquin, who was sent to the West Indies and Central America to collect plants for the garden of the Holy Roman Emperor, Francis I. Upon Jacquin's return, he published his plant discoveries in his *Enumeratio Systematica Plantarum, quas in insulis Caribaeis* (1760). One such discovery was *Epidendrum altissimum,* found in Jamaica (Fig. 3.38).

The small genus *Jacquiniella,* made up of a dozen plants from Latin America, has been named after him by Rudolf Schlechter.

It was the Swedish botanist Olof Swartz who established the genus *Oncidium* in 1800, using the (former) *Epidendrum altissimum* of Jacquin as its type species [21]. The name derives from the Greek *onkos,* referring to the callus growths on the lip of various species. *Oncidiums* are found in all manner of habitats in Latin America, and can be lithophytes, epiphytes, or terrestrial. Of the more than 1400 former species once classified as *Oncidium,* the current number of accepted species

Fig. 3.37 Bear in mind that Mother Nature created these patterns. Clockwise: *Odontoglossum crispum* (*Oncidium alexandrae*); the newly discovered *Odm.* (*Oncidium*) *deburghgraeveanum*, found by Dr. Guido Deburghgraeve at Ecuagenera, the orchid establishment in Ecuador; *Odm.* (*Oncidium*) *blandum*; *Odm.* (*Oncidium*) *cirrhosum* (Images courtesy Steve Beckendorf and Guido Deburghgraeve (*Odm. deburghgraeveanum*))

Fig. 3.38 Jacquin's description of *Epidendrum altissimum* (#15) in his *Enumeratio Systematica Plantarum, quas in insulis Caribaeis,* published in 1760

is 341 [22]. Species have been dispersed to all manner of other genera over the years because of the contentious nature of the defining characteristics of this genus (Fig. 3.39).

Another very important *Oncidium* member classified by Swartz is the epiphytic species *Oncidium leucochilum,* discovered by Ure Skinner in Guatemala. It has imparted its chocolate-like fragrance to the chocolate lover's dream orchid, *Oncidium* Sharry Baby. No instance of human pollination by the act of smelling its fragrance has been recorded, but this could be possible!

Oncidium leucochilum is a typical "dancing lady" type of orchid, with large sprays of flowers

Fig. 3.39 From earliest discovery to a modern hybrid. (L): *Oncidium altissimum,* originally described by Jacquin as an *Epidendrum.* (R): Who can resist the allure of the hybrid *Oncidium* Sharry Baby with its chocolate scent? (Images courtesy (L) Herbarium/ www.plantasyhongos.es/herbarium; (R) Martin Young)

Fig. 3.40 The *Oncidium* chocolate fragrance starts here with *Oncidium leucochilum* (Courtesy José Amorin)

on branched inflorescences up to 3 m (yes, 3 meters) long (Fig. 3.40).

Orchis

We have seen in Chap. 1 how some species members of the genus *Orchis* have played a sexual role in human affairs, even to the present day. This once extensive genus (formerly numbering over 2000 species) has a particularly interesting terrestrial member, *Orchis simia*, commonly known as the "Monkey Orchid." It should not be con-fused with some *Draculas,* which also bear this common name. The orchid is found throughout Europe, the Mediterranean, the Middle East, and North Africa, but in spite of it widespread distribution, it is sparse where it does appear, and over-all its numbers are declining. As the tubers have been used extensively to make salep, this is one reason why the species is endangered in these areas, with loss of habitat being another. In some European countries, the species is now protected (Fig. 3.41).

It just so happens that the Monkey Orchid has a particular sexual relationship with the "Man Orchid," *Orchis anthropophora*, the flowers of which look like little men with their torso-shaped labellum having spindly arms and legs. In a study conducted in France [23], the men and the mon-keys would mingle via a beetle, *Cidnopus pilosus*, which does not discriminate between which spe-cies it pollinates. It goes freely from one to the other, spreading its accumulated pollen masses between monkey and man.

And what does one get when a man is crossed with a monkey? The chimera, *Orchis* × *bergonii*, having just about the appearance one might expect, includ-ing a shortened tail. The frequency of occurrence of this hybrid was found to be closely associated with that of its pollinating beetle (Fig. 3.42).

Fig. 3.41 (L): The delightful *Orchis simia,* or "Monkey Orchid," whose numbers are declining and are in need of our protection. (R): *Orchis anthropophora,* the "Man Orchid," is cross-pollinated with the Monkey Orchid via a beetle (Images courtesy (L) Mark Sewell, (R) Franck Féret)

Fig. 3.42 (L): The Man-Monkey natural orchid hybrid *Orchis × bergonii.* (R): The Lady Monkey Orchid, *Orchis × angusti- crusis* (Images courtesy (L) Mark Sewell, (R) Andrew Porter)

The monkey-business is not solely confined to men, as the natural cross between *Orchis purpurea* (Lady Orchid) and *Orchis simia* yields the hybrid "Lady Monkey Orchid," *Orchis × angusticrusis* (Fig. 3.42) [24].

Paphiopedilum

On anyone's list of orchids with an exotic appearance must be *Paphiopedilums,* or "Lady's Slippers." *Paphs* are unusual in that they have two

Fig. 3.43 The parts of a *Paphiopedilum* showing the fused lateral sepals (synsepal) and staminode (Image of *Paph. gratrixianum* courtesy Sylvia Kappl, illustrated by Katy Metcalf)

anthers attached to the column, unlike most other orchid genera, which simply have one. The column, anthers, and stigma sit behind a curious little shield called the *staminode,* and the distinctive slipper-shaped lip plays an important role in the pollination process, as discussed in Chap. 5. The normally separate two latter sepals are fused into a single entity called a *synsepal.* This is a very unusual-looking orchid indeed and can be found growing terrestrially on forest floors, as a lithophyte, or even as an epiphyte (Fig. 3.43).

Queen Victoria was so enamored with *Paphs* that new ones during her rein were named after her, such as *Paphiopedilum victoria-mariae* and *Paphiopedilum victoria-regina.* The latter is critically endangered and very rare in its habitat of western and northern Sumatra, due to its rampant collection over many decades and the usual destruction of habitat (Fig. 3.44).

Another Paph royalty, *Paphiopedilum rothschildianum,* commonly known as "Rothschild's Slipper Orchid," is found high on the steep slopes of Mount Kinabalu on the island of Borneo. When first discovered in the 1887, it caused an absolute sensation and was considered the "King of Orchids." It was named for the eminent Victorian orchid collector Baron Ferdinand de Rothschild. The flowers' striking appearance is noteworthy for

Fig. 3.44 *Paphiopedilum victoria-mariae,* a slipper truly fit for a Queen. This orchid is endangered and rare in its natural habitat (Image courtesy Orchi/Creative Commons)

their horizontal, highly spotted petals, and they are pollinated by a specific female hoverfly (*Dideopsis aegrota*). The hoverfly deposits its eggs on the staminode, mistaking it for an aphid colony, which is the usual site for the fly to deposit its eggs as food for its larvae. In return, the orchid deposits its eggs on the hoverfly.

Existing numbers on Kinabalu are now considered to be very low as a result of the illicit trade in wild orchids. Fortunately, it can now be widely

Fig. 3.45 The "King of Orchids," *Paphiopedilum rothschildianum,* found in Mount Kinabalu in Borneo. Though scarce in nature, it is available through artificial propagation (Image courtesy Alan Cressler)

Fig. 3.46 *Paphiopedilum sanderianum* was rediscovered in 1978 after it all but disappeared from cultivation (Image courtesy Glen Decker, Piping Rock orchids)

made available by means of artificial propagation (Fig. 3.45).

Another exotic and rare *Paph* from Borneo is the lithophytic *Paphiopedilum sanderianum,* first discovered by accident by Sander's collector J. Förstermann in 1885, while out searching for another *Paph* species. It has arguably the most extraordinary petals of any orchid flower, which can reach nearly a meter in length. However, the orchid proved difficult to grow in cultivation, and all remaining plants effectively disappeared from the scene. The orchid in time took on a mythical status, with its actual existence doubted or thought to be extinct. All that changed when in 1978, the *Paph* was rediscovered by Ivan Nielson growing in the Gunung Mulu National Park of Sarawak, where it is now protected (Fig. 3.46).

Paraphalaenopsis

Although we have seen many beautiful species of *Phalaenopsis* in Chap. 1, in 1963, four species of *Phalaenopsis* from Borneo plus one natural hybrid (*Paraphalaenopsis* × *thorntonii* = *Pps. denevei* × *Pps. serpentilingua*) were reclassified. This was due in part to their long, terete-shaped leaves reaching up to 3 m in length, as is the case for *Paraphalaenopsis labukensis* [25] (Fig. 3.47).

Fig. 3.47 The large (6 cm/2.5 in) scented flowers of *Paraphalaenopsis labukensis* from Borneo has terete-shaped leaves up to 3 m (10 ft) in length (Image Steve Beckendorf)

Phragmipedium

There are other slipper orchids of related genera, the most common being *Phragmipediums* and *Cypripediums* (discussed earlier), which have a very similar appearance and structure to *Paphiopedilums* but certain genetic and morphological differences. In *Phragmipediums,* the lateral sepals are fused together to form a synsepal, as with the *Paphiopedilums,* but the edges of the lip are rolled inward, unlike *Paphs*. More than 20 species grow throughout the Americas. They flower in the spring, and the plant can be in flower for more than 6 months with multiple flowers appearing along a stem.

Relatively recently, in 1981 Elizabeth Locke Besse made a discovery in Peru of the terrestrial species *Phragmipedium besseae*. This orchid is brilliant orange in color and has breathed new life into this genus, with many new hybrids created from it. Even more spectacular is the somewhat notorious new discovery, also in Peru, of *Phragmipedium kovachii*, named after James Kovach. Kovach was involved in illegally transport-ing the orchid to the U.S.A., for which he was prosecuted and fined. Subsequently, its native habitat was pillaged of many thousands of plants, and naturally, much controversy attended the naming of this orchid after the person who played such an infamous role. On the other hand, many other historic orchid species have been named after an individual who was personally involved in the pillaging of thousands of plants in the wild, so the name stands. Even the CITIES Treaty has not been immune from criticism in this unsavory saga. As the orchid is now available through artificial propagation, there is no longer any need to procure it from the wild [26].

The magnificent new orchid is unusual in some respects. The plant grows in its native cloud forest habitat on a medium of crushed limestone (calcium carbonate), and a small amount of organic matter, which provides a slightly alkaline environment (pH 7.9). The flower can grow to 18 cm (7 in) or more across and is of a delicate mauve coloring (Fig. 3.48).

A more *Paph*-like *Phragmipedium* would be the beautiful spotted *Phragmipedium pearcei,* also from Ecuador and Peru (Fig. 3.49).

Fig. 3.48 The recently discovered (2002) *Phragmipedium kovachii* (L) is up to 18 cm (7 in) across. (R): the considerably smaller (4 cm to 6.5 cm) *Phragmipedium besseae*. Both have been used extensively for hybridizing (Images courtesy (L) Guido Deburghgraeve and (R) Roberta Fox)

Fig. 3.49 The striking *Phragmipedium pearcei* (Image courtesy Dalton Holland Baptista)

Pleurothallis

This was formerly a very large genus that was decimated in recent years, not by plunder in the wild, but rather by lab DNA analysis and subsequent species reclassifications that left the genus *Pleurothallis* with little over half its former members (557 as of 2017). The diaspora has seen many of the genus's species end up in other genera, including *Acianthera, Anathallis, Specklinia*, and *Stelis.*

The name derives from *pleuron*, Greek for "rib," and *thallos,* meaning "shoot," in reference to the ribbed stems of various species. The genus was given this title in 1813 by the eminent Scottish botanist Robert Brown (the same man for whom Brownian motion is named), who was the first to observe it. Some species have the odd characteristic of the flowers lying on top of the leaf as in *Lepanthes*. These small plants with their small flowers are often found in cloud forests of Latin America (Fig. 3.50).

Polycycnis

The aptly named small genus *Polycycnis, poly* being Greek for "many" and *kyknos* for "swan," is found in Central and South America, often with several flying swans on each inflorescence. One unusual feature is its long, thin column, arching high over a hairy lip.

The flowers are scented and attract male euglossine bees (commonly known as "orchid bees") that climb onto the lip and scrape with their forelegs to gather its fragrant compounds. In doing so, the bee's weight compresses the hinged lip and column downward so that when the bee flies off, it hits the anther and the pollinarium becomes stuck to its thorax [27] (Fig. 3.51).

Ponthieva

Henri de Ponthieu was a London merchant and amateur botanist who, during his time in the West Indies, sent specimens of plants and fish to the Royal Society via famed British naturalist Joseph Banks. His name was awarded to a genus of about 30 species of generally terrestrial orchids from the Americas and West Indies, by the same Robert Brown who named the genus *Pleurothallis*. The flowers are non-resupinate (so the flowers look upside down but are actually right-side up). The spooky appearance of some species gives the genus the common name "Shadow Witch (Fig. 3.52)."

Pollination of this genus has not been widely studied, but Dressler found that the labellum secretes an oil and suggested it could be collected by a type of bee (*anthrophorid*) as a food source for its larvae [28]. This needs further investigation.

Porroglossum

About 50 or so species of epiphytic and terrestrial orchids comprise this genus from the Andes region of South America. The name derives from the Greek word for "forward" and *glossa* for "tongue," referring to the jutting out position of the lip. The genus was created by German botanist and orchid taxonomist Rudolf Schlechter (1872-1925), who also created several other new genera, including *Holcoglossum* and *Neomoorea*. Some *Porroglossum* species were formerly classified by Reichenbach as *Masdevallia.*

Fig. 3.50 Clockwise from top: *Pleurothallis silverstonei; Pths. nuda, Pths. caprina; Pths. fustifera* (Images courtesy Wiel Driessen)

The lip is unusual in that it is hinged, and when an insect lands on it, the lip suddenly snaps shut, trapping the insect against the pollinia. After a period of approximately half an hour, the lip returns to its normal position and releases the insect. Trapping a pollinating insect for an extended of time makes for a more likely transfer of pollen either to or from the insect (Fig. 3.53).

Prosthechea

Prosthechea cochleata, or "Cockleshell Orchid," is unusual in that the lip is not only broad and looks like a cockleshell, but the flowers are non-resupinate, giving the flower the appearance of a 5-legged octopus. An epiphyte, it is found in Central America even up to southern Florida,

Fig. 3.51 *Polycycnis escobariana* (named after Colombian orchid enthusiast Rodrigo Escobar) with its slender, high-arching column and hairy lip (Image courtesy Alan Cressler)

and it is the national flower of Belize. The pseudobulbs contain a sticky substance that has been used as an adhesive. When the author first bought this orchid, it was called *Encyclia cochleata,* but since 1997, it has been reclassified under the new genus *Prosthechea* (Greek word *prostheke,* meaning "addendum"), which refers to an appendage on the back of the column. The flowers are said to smell like honey.

Many species of this genus have flowers that are borne on the inflorescence in a non-resupinate manner, whereas others display in the resupinate form (Fig. 3.54).

Restrepia

This is another genus on everyone's list of exotic orchids. The genus of these small compact orchids is named after José Manuel Restrepo, a Colombian politician who had an interest in

flora. The dorsal sepal and petals are slender and antenna-like, with the lateral sepals fused to form a curved synsepal. Its habitat is the rain and cloud forests of Central and South America, with *Restrepia antennifera* found at elevations up to 3500 m (11, 500 ft), which was one of the specimens brought back by von Humboldt and Bonpland. Though epiphytic, the species lack pseudobulbs, and usually multiple blooms are born simultaneously but singly on thread-like stems. They have a very endearing alien insect appearance in gorgeous color combinations of spots and stripes (Fig. 3.55).

Rhizanthella

Of all the evolutionary pathways that have been explored by orchids over millions of years, none is more bizarre than the three species of this genus that grow underground in Australia. The most studied species, *Rhizanthella gardneri* – only discovered in 1928 in Western Australia – consists of a tuber that feeds off the fungi (myco-heterotrophy) associated with the roots of the broom bush (*Melaleuca uncinata*). The plant lives a completely subterranean existence, even flowering below ground! The chloroplast cells that normally perform the function of photosynthesis have been retained, however, they have lost nearly 70% of their genes and now only code for a few essential proteins that are necessary for the life of an underground parasite [29] (Fig. 3.56).

Another Australian endangered underground orchid species is *Rhizanthella slateri*, but in this case, the flowers can just be seen at the surface of the eucalypt forest floor where it has been found to grow.

Rhynchostele

Although this genus was created by Reichenbach in 1852 with the type species *Rhynchostele pygmaea*, various members of the *Rhynchostele* genus look like *Odontoglossums,* and for many

Fig. 3.52 Clockwise: *Ponthieva brenesii; Ptva. formosa* (there have been some taxonomic issues regarding these two species as well as *Ptva. maculata*); *Ptva. racemosa; Ptva. tunguraguae* (Images courtesy (clockwise): Daniel McLaren, Daniel McLaren, Christian Furtwängler, Wiel Driessen)

years, they were classified as such. Then they became *Cymbiglossum,* and subsequently, *Lemboglossum.* Some were also *Amparoa,* others *Mesoglossum.* Finally in 1993, they transferred to *Rhynchostele* [30].

This genus of 20 species, including two natural hybrids, is found throughout Mexico, Central America, and Venezuela. The circular markings around the column of some species could serve as a useful target for pollinators (Fig. 3.57).

Fig. 3.53 Clockwise from top: The waiting lips of *Porroglossum amethystinum; Prgm. nutibara;* and *Prgm. teaguei)*Images courtesy Joost Riksen and Wiel Driessen (*Prgm.teaguei*))

Scaphosepalum

An unusual name aptly describes this unusual Latin American genus consisting of more than 40 highly bizarre-looking species that can be either epiphytic, lithophytic, or terrestrial. The name derives from the Greek *skaphos,* meaning "hollowed out," and *sepalum* for "sepal," owing to the bowl shape of the sepals. Species flowers are non-resupinate and have been found to possess osmophores at the tips of the sepals (dorsal, lateral, or both), which exude such charming fragrances described as "rancid," "rotting fish," or the most innocuous "non-detectable to humans." With such attractions, the pollinators are suspected to be flies [31] (Fig. 3.58).

Zootrophion

This exotic-sounding genus was named by the famed orchid botanist Carlyle Luer in 1982, after the Greek word for "menagerie," owing to the animal head-like appearance of the flowers. There are about 24 epiphytic species growing in the cloud forests of the Caribbean, Central and South

Fig. 3.54 Clockwise from top left: *Prosthechea vitellina; Psh. prismatocarpa; Psh. cochleata; Psh. fragrans.* The first two are resupinate and the second two are non-resupinate (Images courtesy Ramūnas Pileičikas, John Varigos, Ruud de Block, Mabelín Santos)

America, with many species transferred from the now defunct genus, *Cryptotheanthus.* The flowers are unusual in that lateral sepals and dorsal sepal are all joined at their tips, forming oval slits that expose the interior, which house the small petals and a trilobed labellum. This gives the orchid flower(s) born on a single stem the appearance of a piece of ripe fruit that is splitting open. They are thought to be pollinated by small flies (Fig. 3.59).

Fig. 3.55 Clockwise: *Restrepia sanguinea; Rstp. antennifera* ssp. hemsleyana; *Rstp. driessenii*, found in the Tovar region of Venezuela by orchid aficionado Wiel Driessen; *Rstp. metae; Rstp. chameleon; Rstp.chrysoglossa; Rstp. muscifera; Rstp. nittiorhyncha* (All images courtesy Wiel Driessen)

Fig. 3.56 The bracts of the *Rhizanthella gardneri* house the approximately 150 flowers of this subterranean orchid. The dirt has been brushed away to expose the flowers (Photo courtesy Justin Brown)

Fig. 3.57 Clockwise from left: *Rhynchostele cordata; Rst. maculata; Rst. cervantesii*. The last is named for Vicente de Cervantes (1755 – 1829), a Mexican botanist, and not for the great Spanish author of *Don Quixote*, Miguel de Cervantes (All images courtesy Norbert Dank/www.flickr.com/photos/nurelias)

Fig. 3.58 The bizarre world of *Scaphosepalum*. Clockwise from left: *Scaphosepalum beluosum; Scaph. breve; Scaph. fimbriatum; Scaph. swertiifolium* (Images courtesy Wiel Driessen)

Fig. 3.59 The flowers of *Zootrophion argus* resemble ripe fruit splitting open (Image courtesy Wiel Driessen)

References

1. Antitumor and immunostimulatory effects of *Anoectochilus formosanus* Hayata, *The Free Library*, Urban & Fischer Verlag, 2016: https://www.thefreelibrary.com/Antitumor+and+immunostimulating+effects+of+Anoectochilus+formosanus...-a0147344278/. The entire plant was ground up with added water and the mixture filtered and injected into the tumor bearing mice. 'This study suggests that the antitumor activity of *A. formosanus* may be associated with its potent immunostimulating effect.'

2. C.J. Bulpitt: https://academic.oup.com/qjmed/article/98/9/625/1547881/The-uses-and-misuses-of-orchids-in-medicine

3. André Schuiteman, *et al.*, Nocturne for an unknown pollinator: first description of a night flowering orchid (*Bulbophyllum nocturnum*), *Botanical J. of the Linnean Soc.*, 167 (3), 344–350, 2011.

4. D. Zhang, *et al.*, Dactylorhin B reduces toxic effects of β-amyloid fragment (25–35) on neuron cells and isolated rat brain mitochondria, *Naunyn-Schmiedeberg's Arch. Pharm.*, 374 (2), 2006, 117–125.

5. *Nova Acta Regiae Soc. Sci. Upsal.* 6, p.70, 1799.

6. M. Sugahara, *et al.*, Oriental orchid (*Cymbidium floribundum*) attracts the Japanese honeybee (*Apis cerana japonica*) with a mixture of 3-hydroxyoctanoic acid and 10-hydroxy-(E)-2-decenoic acid, *Zoolog. Sci.* 30(2), 99–105, 2013.

7. Jun-Pin Xu, *Cancer Inhibitors from Natural Chinese Medicines*, CRC Press, 2016.

8. Z. Yu, *et al.*, *Dendrobium chrysotoxum* Lindl. Alleviates Diabetic Retinopathy by Preventing Retinal Inflammation and Tight Junction Protein Decrease, *J. Diabetes Res.* Vol. 2015, 2015, 10pp.

9. N.A. Van Der Cingel, *An Atlas of Orchid Pollination: European Orchids*, CRC Press, 2001, p.177.

10. M.A. Clements, *et al.*, *Winika*, a new monotypic genus for the New Zealand orchid previously known as *Dendrobium cunninghamii* Lindl, *The Orchadian* 12, 214–215, 1997.

11. J. J. Sadler, *et al.*, Fragrance composition of *Dendrophylax lindenii* (Orchidaceae) using a novel technique applied in situ, *Eur. J. Enviro. Sci.*, 1 (2), 137–141, 2011.

12. L. Endara, *et al.*, Lord of the Flies: Pollination of Dracula orchids, *Lankesteriana* 10 (1), 1–10, 2010.

13. F. Pinheiro, *et al.*, Phylogenetic relationships and infrageneric classification of *Epidendrum* subgenus *Amphiglottium* (Laeliinea, Orchidaceae), *Plant Syst. Evol.*, 283 (3), 165–177, 2009.

14. A.H. Kennedy and L.E. Watson, *Syst. Bot.* 35(1), 74, 2010.

15. M. Blanco and G. Carnevali, *Lankesteriana*, 7, 527, 2007.

16. R.B. Singer, *et al.*, *Brasiliorchis*: A New Genus for the *Maxillaria picta* Alliance (Orchidaceae, Maxillariinae), *Novon*, 17 (1), 91–99, 2007.

17. R.B. Singer, *et al.*, Proposal to conserve the name *Brasiliorchis* against *Bolbidium* (Orchidaceae), *Taxon*, 60 (6), 1775–1775, 2011.

18. Godefroy-Lebeuf, *L'Orchidophile: Journal des Amateurs D'Orchidee*, 9, 63–64, 1889.

19. G.C.K. Dunsterville, and L. A. Garay, *Venezuelan Orchids Illustrated,* 6, Harvard University, Cambridge, Mass., 1976.

20. S. Dalström and G. Merino, A new species of *Odontoglossum* (Orchidaceae: Oncidiinae) from Ecuador, *Lankesteriana* 9 (3), 505–508, 2010.

21. *Kongl. Vetensk. Acad. Nya Handl.* 21, p.239, 1800.

22. www.theplantlist.org.

23. B. Schatz, Fine scale distribution of pollinator explains the occurrence of the natural orchid hybrid ×*Orchis bergonii*, *Ecoscience* 13(1):111-118. 2006.

24. R.M. Bateman, *et al.*, Morphometric and population genetic analyses elucidate the origin, evolutionary significance and conservation implications of *Orchis* × *angusticruris (O. purpurea* × *O. simia)*, a hybrid orchid new to Britain, *Bot. J. Linnean Soc.*, 157, 687-711, 2008.

25. A.D. Hawkes, *Orquídea* (Rio de Janeiro) 25, 212, 1963.

26. Staff writer, Craig Pittman, of the *Tampa Bay Times* has written a wonderful book covering the entire intriguing, but infamous tale of *Phragmipedium kovachii*, *The Scent of Scandal: Greed, Betrayal, and the World's Most Beautiful Orchid*, Univ. of Florida Press, 2012.

27. The Orchid Column, *Notes from the Fuqua Orchid Center*, B. Brinkman, Oct.21, 2013.

28. R. Dressler, *Phylogeny and Classification of the Orchid Family*, Cambridge University Press, 1993, p.123.

29. E. Delannoy, *et al.*, Rampant gene loss in the underground orchid *Rhizanthella gardneri* highlights evolutionary constraints on plastid genomes, *Mol. Biol. Evol.*, 28 (7) 2077–2086, 2011. AND: http://www.news.uwa.edu.au/201102073251/research/was-incredible-underground-orchid

30. Soto Arenas & Salazar, *Orquídea* (Mexico City), n.s., 13, 146–151, 1993.

31. A.M. Pridgeon and W.L. Stern, Osmophores of *Scaphosepalum* (Orchidaceae), *Bot. Gaz,* 146, No. 1, 1985, 115–123.

Deviant Sex

4

In my examination of orchids, hardly any fact has struck me so much as the endless diversities of structure – the prodigality of resources – for gaining the very same end, namely, the fertilization of one flower by pollen from another plant.

Charles Darwin

Orchids have evolved a multitude of seemingly bizarre contrivances to get themselves pollinated. Indeed, Charles Darwin made painstakingly detailed studies of orchid reproduction processes, published in his 1862 book *The Various Contrivances by Which Orchids Are Fertilized by Insects.* This book was meant to provide further evidence in support of his revolutionary evolutionary theory, expounded in his great work *On The Origin of Species by Means of Natural Selection* published 3 years earlier. Writing to publisher John Murray, Darwin said of the book, "I think this little volume will do good to the 'Origin', as it will show that I have worked hard at details." Indeed he did, but his endeavors did not convince the theologically minded, as a review in the *Literary Churchman* concluded that the book was in essence saying, "Oh Lord, how manifold are Thy works!"

Some of the discussion concerning the sexual lures employed by orchids discussed herein actually go back to the original studies by Darwin himself (Fig. 4.1).

Darwin and His Moth

One of the most unusual flowers in the orchid world is *Angraecum sesquipedale,* also known as the "Comet Orchid" due to its an exceptionally long, thin tube (spur) protruding from the back of the lip. Flowers of other orchid species also have a nectar-containing spur (e.g. *Neofinetia, Habenaria*), but the spur of this species can reach up to 30–35 cm (12–13 in – figures vary)! In addition, the flower can span up to 22 cm (9 in) across. This gives rise to the Latin name *sesqui,* which means "one-and-a-half times" (like sesquicentenary – the 150th anniversary), and *pedalis,* which translates as "measuring a foot," in reference to the one-and-a-half foot *Angraecum* flower, measured from the bottom of the spur to the tip of the dorsal sepal. This very sturdy monopodial plant is enormous too and can grow over a meter tall. At the orchid nursery where the author works part-time, there are two such behemoths, and sitting amongst them, one is immediately transported to the forests of Madagascar.

This strongly scented species has a nectar reward near the bottom of the spur (or nectary), just waiting for any pollinator capable of reaching it. In *Contrivances* (1862), Darwin writes, "In several flowers sent me by Mr. Bateman [English horticulturalist James Bateman] I found the nectaries eleven and a half inches long, with only the lower inch and a half filled with very sweet nectar." But what creature has the facility to pollinate such a flower?

Darwin suspected that it must be a sphinx moth with an unusually long proboscis (tubular sucking tongue): "…Our English sphinxes have probosces as long as their bodies: but in Madagascar there must be moths with probosces capable of extension to a length of between 10 and 11 inches!" To substantiate his claim, Darwin experimented using a thin tube to probe the depths of the spur. Sure enough, when withdrawing the tube, the pollen grains would stick to the tube. When reinserted into another such nectary, the pollen grains would be subsequently deposited onto the stigma, thereby pollinating the flower (Fig. 4.2).

© Springer International Publishing AG 2018
J.L. Schiff, *Rare and Exotic Orchids*, https://doi.org/10.1007/978-3-319-70034-2_4

Fig. 4.1 The father of evolutionary theory, Charles Darwin, in 1874

Fig. 4.2 The famous *Angraecum sesquipedale* has such long spurs that only a moth with a sufficiently long proboscis can reach the nectar held at the bottom. Note also the protruding landing pad labellum (Image courtesy Olivier Reihes)

Darwin went on to explain how such a long nectary could have evolved. Through natural variation, there would be moths with a longer proboscis than average. It would be those *Angraecum* flowers with the longest nectary that "compelled the moths to insert their proboscis up to the very base, [that] would be fertilized. These plants would yield most seed, and the seedlings would generally inherit longer nectaries; and so it would be in successive generations of the plant and moth. Thus it would appear that there has been a race in gaining length between the nectary of the Angræcum, and the proboscis of certain moths; but the Angræcum has triumphed, for it flourishes and abounds in the forests of Madagascar, and still troubles each moth to insert its proboscis as far as possible in order to drain the last drop of nectar [1]."

But did this happen in real life? At the time, there was no known insect that was capable of reaching the nectar, so Darwin had to postulate one based on his evolutionary theory: a moth with a foot-long proboscis. Sadly, Darwin did not live to see the discovery in 1903 of just such a moth, aptly name *Xanthopan morganii praedicta* and referred to as Darwin's moth. No doubt he was smiling from his grave at this astute bit of confirmed scientific deduction (Fig. 4.3).

It should be mentioned that the *Angraecum sesquipedale* scent is most prominent at night, which matches the foraging habits of the *Xanthopan morganii praedicta,* and it is also thought that some of the 210 (!) volatile compounds could be important attractants for this particular pollinator [2]. Different parts of the flower are primarily responsible for portions of the volatile chemical spectrum (see Appendix II) (Fig. 4.4).

There are various related tropical genera, such as the African *Aerangis*, whose species members also possess long spurs of varying lengths, such that the length of the spur is in direct accordance with the length of the pollinating hawkmoth's proboscis. Moreover, for some *Aerangis* nectaries that were nearly full, it was found that the sugar concentration of the nectar increased with the depth of the spur, thus encouraging the hawkmoth to probe deeper. This is another evolutionary twist to first lure in the hawkmoth, then have it plunge deep to ensure pollination [3].

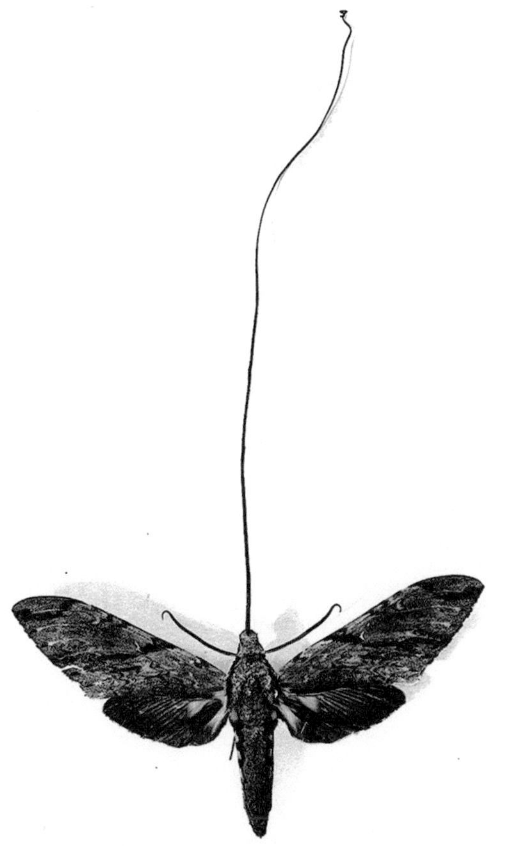

Fig. 4.3 Darwin's moth, found in Madagascar and East Africa has a footlong proboscis. Normally, it is rolled up and tucked away, but it extends for probing and feeding (Image courtesy Natural History Museum, London)

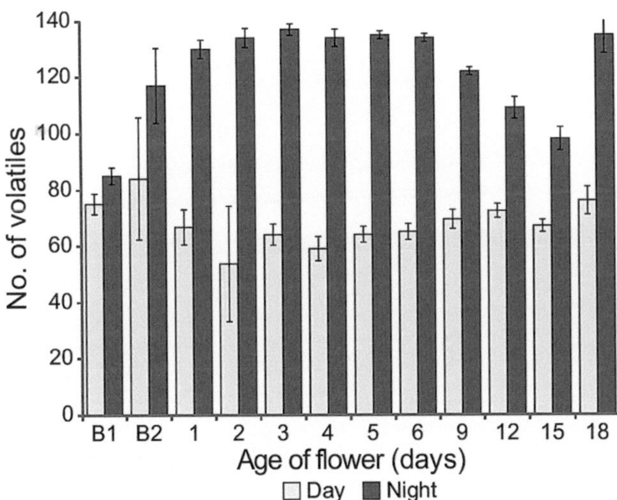

Fig. 4.4 The staggering number of registered volatile compounds from flower samples of *Angraecum sesquipedale* at day and night during the entire flowering period (From L.J. Nielsen and B.L. Møller, *ibid.* See also Appendix II for a list of the organic compounds)

Elaborating on Darwin's explanation of how a long nectar spur could evolve, co-evolutionist Alfred Russel Wallace maintained in an 1867 paper that if the orchid spur is shorter than average, then the (average) pollinating moth would have no trouble reaching the nectar without having to brush against the pollinia., and consequently, these orchids would risk not being fertilized and have a larger reproductive disadvantage. On the other hand, an orchid with a longer-than-average spur would be pollinated more frequently as moths struggle to secure the out-of-reach nectar, resulting in the pollinia being attached to them. This would result in the nectar spur becoming longer over time, as Darwin suggested. Similarly, "the moth would also be affected, for those with the longest proboscis would get the most food, would be the strongest and most vigorous, would visit and fertilize the greatest number of flowers and would leave the largest number of descendants … there would in each generation be on average an increase in the length of the nectaries, and also an average increase in the length of the proboscis of the moths …" [4].

In an added evolutionary twist, some *Angraecum* orchids such as *Angraecum arachnites*, also found in Madagascar, have spirally twisted spurs that encourage the pollinating moth to go ever deeper and spend longer time in close contact with the orchid flower (Fig. 4.5).

This interactive evolution between two species, as exemplified by the Darwin and Wallace model, has been termed "coevolution." As convincing as this model sounds, there is an alternative explanation known as the "pollinator shift model" that attributes the lengthening of the moth's proboscis to the need for the moth to hover and move about at a distance from the flower to avoid predators [5].

It should be mentioned that either way, the well-established biological concept of coevolution is not affected by the validity of the preceding Darwin and Wallace nectary/proboscis explanation, and a further instance of it will be discussed in the next chapter. But clearly, further research needs to be done regarding the evolution of long nectar spurs and the corresponding lengthy proboscis of its pollinator (Fig. 4.6).

Fig. 4.5 The illustration by Thomas W. Wood that accompanied Alfred Russel Wallace's 1867 article on evolution. It depicts a (hypothetical) "Sphinx moth fertilizing *Angraecum sesquipedale* in the forests of Madagascar"

Fig. 4.6 *Aerangis ellisii* from Madagascar has long curved nectar spurs on each flower (Image courtesy John Varigos)

Staying in this geographical region, another very unusual instance of pollination is found on the island of Réunion 480 km east of Madagascar. Here, hawkmoths are absent, and so one would not expect any *Angraecum* species there to have a long, spurred nectary. And indeed the case is quite the opposite.

In this instance, there is the endemic species *Angraecum cadetii*, which has a rather small whitish-green flower (1.25 cm) and a very short nectary (~6 mm). The headroom between the lip and pollinarium leading to the nectary at the back of the flower is just a whisker smaller than the head of a raspy cricket, which was discovered to be the pollinator. When the cricket forces its head into the nectary opening for its sugary reward, the orchid is also rewarded by having its pollen deposited on the head of the perpetrator. The next time the cricket sticks its head into another *Angraecum cadetii* for its nectar, the pollen is deposited on the orchid's stigma and the pollination cycle is complete. Interestingly, although the orchid is also visited by birds and other insects, none of them were found to remove pollinia (Fig. 4.7).

Crickets normally eat other plants and insects, but possibly due to lack of other insects on the island, a highly unorthodox symbiotic relationship has developed between orchid and cricket. Indeed, this is the first recorded instance of a cricket pollinating a flower of any kind, and the cricket itself is a new species [6].

Fig. 4.7 The raspy cricket *Glomeremus orchidophiles* investigates an *Angraecum cadetii* on the island of Réunion. The cricket is a new species (Image courtesy Sylvain Hugel)

The Paph Way

One way to ensure fertilization is to capture a pollinator for a brief period of time, as we have seen in the genus *Porroglossum.* In *Paphiopedilums,* however, the sexual apparatus is arranged a little differently than in other orchids. In fact there are two anthers attached to the column, with the anthers and stigma located behind the shield-like staminode. Then there is the unusual labellum, shaped like a slipper with very slippery internal sides. An insect may land on the staminode and either fall or fly into the pouch of the lip. It is unable to climb its way out except at the rear wall of the pouch, where there are very tiny upwardly pointing hairs for it to grasp. The insect can go either left or right, but in both cases, it will either pick up a pollen packet or brush against the stigmatic surface, pointing downwards as in (Fig. 4.8).

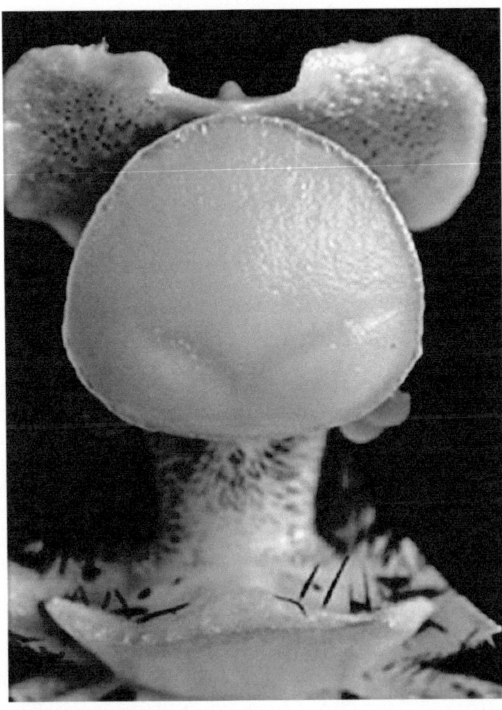

Fig. 4.8 The insect's view of its escape from the pouch of a *Paphiopedilum,* showing the disk shape of the stigmatic surface and the two anthers on either side of the column. The top structure is the staminode seen from below (Courtesy Brian Johnston)

Visual Deception

About one-third of all orchid species do not produce any nectar reward for pollinators, which means that they have to be a bit more cunning and deceive potential pollinators either visually, by scent, or both. But the evolutionary success of non-rewarding orchid flowers and their overrepresentation in orchids is problematic, as many have low rates of visiting pollinators. Nevertheless, deception as practiced by orchids is what biologists call an *evolutionarily stable strategy* (ESS), which means that it is resistant in the game of life to other competing strategies, such as offering a nectar reward. Incidentally, the concept of an ESS is a biology-oriented refinement of the notion of *Nash Equilibrium* in Economics, for which John Nash won the Nobel Prize in 1994 and of which the fine movie, *A Beautiful Mind,* was made.

Why this should be so, given the fact that providing no reward is "likely to have been the ancestral condition," [7] and providing a nectar reward is a "derived condition," is still a subject for scientific investigation.

One such example of non-nectar rewarding, visually deceptive orchids is the very exotic-looking members of the *Brassia* genus. Named after the nineteenth century British botanical illustrator William Brass, this Central and South American native has flowers with very elongated petals and sepals, which give it the name "Spider Orchid." Each inflorescence can produce up to a dozen fragrant blooms (Fig. 4.9).

The spider-like appearance of the flowers actually serves a biological purpose. Two types of spider-wasps, *Pepsis* and *Campsomeris,* are deceived into thinking the flower is an actual spider. They will attempt to sting the flower on the lip in order to paralyze it and take it back to their nest as food for their larvae. Unfortunately, in the wasp's efforts to grasp the stung "spider," it only succeeds in having the pollinarium stuck to itself, and off it will go to try this again with another *Brassia,* which leads to pollination. An orchid plant with no brain has once again outsmarted an insect that has one, through the mechanism of evolution.

Fig. 4.9 The ever-popular *Brassia verrucosa* has spider-like flowers that can grow up to 15 cm (6 in) (Image courtesy John Varigos)

Another very simple means of deceiving without nectar production is to simply mimic the appearance of nearby flowers of species that do produce nectar, a strategy known as *floral Batesian mimicry*. Evidence for this has been discovered with the orchid *Disa ferruginea,* found in the Cape Province of South Africa. It is pollinated solely by the butterfly *Aeropetes tulbaghia*, yet the *Disa* offers the butterfly no nectar reward in return (Fig. 4.10).

Curiously, there are two color forms of *Disa ferruginea*: red in its western habitat and orange in its eastern parts. It is more than mere coincidence that the red-flowered *Tritoniopsis triticea* is found in its western habitat of the *Disa,* and the orange flowered *Kniphofia uvaria* is found in its eastern habitat. The flowers of both of these (non-orchid) plants offer a nectar reward to pollinators. A careful study has revealed that in each habitat, the *Disa* has specifically adapted its own color to that of either the red or orange nectar-rewarding flower species [8]. This color transition of the *Disa* is mediated by the butterfly that associates the red/orange floral color in the respective habitats with a nectar offering, and in this deceptive manner, the orchid gets itself pollinated. Thus, we not only have an instance of floral mimicry on the part of the *Disa*, but also floral variation driven by its pollinator (Fig. 4.11).

Other food-deceptive or unrewarding orchid species that have been tested for similarity in color

Fig. 4.10 The South African butterfly *Aeropetes tulbaghia,* also known as "Table Mountain Beauty" or "Mountain Pride," has a strong red/orange color affinity. It pollinates the orchid *Disa ferruginea* as a consequence of floral mimicry, which it is in the process of doing here. Note the pollinia on its proboscis (Image courtesy S.D. Johnson)

with co-occurring plant species that do provide nectar rewards for their pollinators can be found in Appendix III.

Another type of deception is achieved by some species of *Oncidium*. Female *Centris* bees are known to gather around certain flower clusters of *Oncidium*, and the male bees are keen to include such clusters in their territory as a means of finding such females. In a breeze, the *Oncidium* flowers move enough to look like an insect in flight to these territorial males, and this arouses their aggression in a process known as *pseudoantagonism*.

There is no reward on offer, but the bee attack is sufficient for the *Oncidium* flower to have its pollinia attached to the bee's head, and pollination is complete when the bee attacks another flower. It has been reported that the territoriality of the bees is "so pronounced, that captured bees return to the defense of their territory upon release [9] (Fig. 4.12)."

Fig. 4.11 (L) The red and orange varieties of *Disa ferruginea*. (R): *Tritoniopsis triticea, and Kniphofia uvaria,* the plants being imitated (Images courtesy (clockwise): Mariana Delport, H.G. Robertson, Iziko Museums of South Africa, Joel Schiff, Marland Holderness)

As we have seen, orchids can be very cunning in their means of deception, and none more so than the very rare and endangered species, *Cypripedium fargesii,* found in high, rocky, mountainous parts of Southwestern China. But this compactly growing baby slipper orchid has a devious trick up its sleeve, or rather, up its leaves. These are spotted in a random fashion as if they have a fungal infection, a despairing

Fig. 4.12 The Central American *Oncidium planilabre* is pollinated by a male *Centris* bee through the mechanism of pseudoantagonism (Image courtesy Orchi/Wiki Commons)

Fig. 4.13 The fake fungal spots of the *Cypripedium fargesii* from Southwestern China (Image courtesy Piotr Markiewicz)

Fig. 4.14 *Holcoglossum amesianum* looks innocent enough but has an unusual trick up its sleeve when it comes to pollination (Image courtesy Martin Guenther)

If All Else Fails: Do It Yourself

In general, orchids are pollinated by birds and insects, and there is inhibiting evolutionary pressure against self-pollination, although in some cases, this does occur. In an extraordinary feat of acrobatics, the epiphytic *Holcoglossum amesianum* from southern China will deliberately pollinate itself! This could be very handy when pollinators are scarce and wind is not available (Fig. 4.14).

Figure 4.15 shows the various stages occurring during this process. In the arena of the orchid's reproductive system (a), the following unfolds: In the first instance, the anther cap pops open (c) exposing the pollinia. The two pollen grains attached to the stipe unfurl from their folded-up positions (d), and then the flexible stipe bends around the rostellum, the barrier that separates the male and female parts (e), in order to then travel upward against the force of gravity to achieve self-fertilization (f) and (g) [10]. A curious feature of this flower is that it also has a rostellum to keep the male and female organs separated.

sight well known to any orchid grower. But of course, the leaves have no such affliction. As it turns out, this is a sight beloved by the hoverfly *Cheilosia lucida*, which is attracted to the black mold spots and lured by the faint unpleasant scent of the fungus the orchid is pretending to have (*Cladosporium*). The hoverfly attempts to feed on the fungus spores, and in the process may end up in the spotted labellum of the flower. In its attempt at escape, a pollen mass is deposited on the insect's thorax. The next time the fly is lured to another *Cypripedium fargesii* for the same purpose, pollination can be achieved (Fig. 4.13).

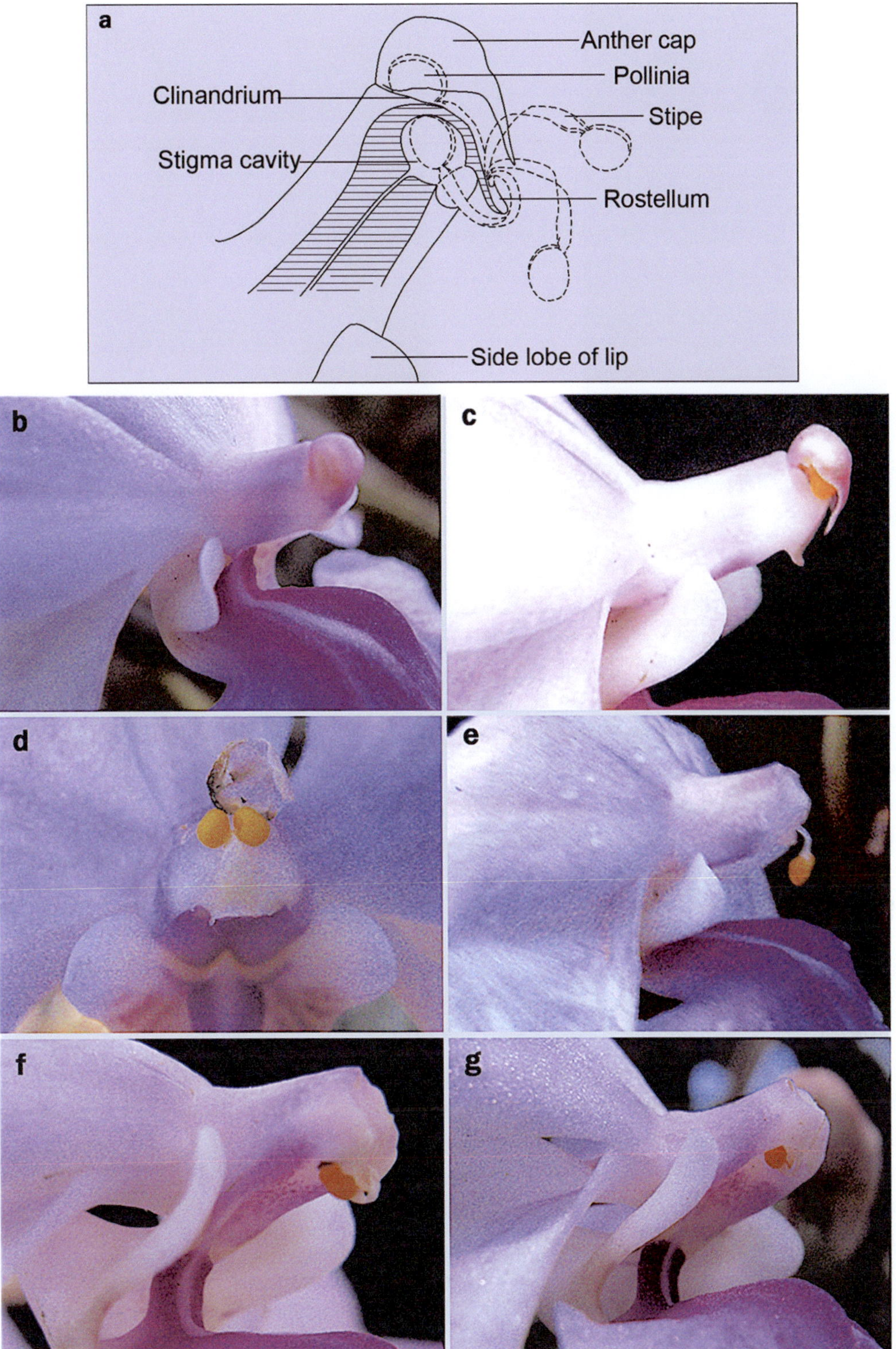

Fig. 4.15 (Top) The anatomical structure of the *Holcoglossum amesianum* indicates the path taken by the pollinia in order to reach the stigmatic cavity and self-pollinate the flower. (Bottom) This close-up view of the sexual apparatus of the *Holcoglossum amesianum* shows the remarkable movements of the stipe and pollinia to reach the stigmatic cavity, all achieved by the flower's own initiative. Pre-pollination state (**b**); self-pollination initiated when anther cap pops open (**c**); pollinia exposed (**d**); flexible stipe bends around the rostellum (**e**); pollinia approach the stigma (**f**); pollinia attach to stigma (**g**). (Images courtesy Zhong-Jian Liu)

Fig. 4.16 The handsome *Paphiopedilum parishii*, which fertilizes itself (Image courtesy Ramūnas Pileičikas)

Fig. 4.17 *Ophrys apifera* with its dangling pollinia requires only a slight breeze to self-pollinate, yet it has also evolved to resemble a bee (Image courtesy Bernard Dupont)

Another do-it-yourselfer is the species *Paphiopedilum parishii,* which has evolved a technique much less elaborate than *Holcoglossum amesianum.* When the flower is ready, the anther and pollinia become liquefied and the droplet slides onto the stigma, simple as that [11]. The authors comment that, "It is a successful evolution indicating that the reproductive assurance of selfing has greater function than inbreeding depression …" Again, this behavior has likely arisen due to a lack of other available pollinators, and it allows for the orchid plant's continued survival (Fig. 4.16).

The simplest mechanism of all self-pollinators belongs to that of the European terrestrial species *Ophrys apifera.* This is another bee-like orchid, but the flowers have no need of a bee for pollination. Instead, pollination is easily achieved by the wind, as the pollinia dangle on thin flexible stipes in front of the stigmatic surface, and a breeze is all it takes for the plant to fertilize itself. This behavior was noted by botanist Robert Brown and also by Darwin. However, Darwin was perplexed by this orchid, because although he could see that it was self-fertilized, it was also perfectly designed to be insect pollinated with its bee-like appearance. He observed, "Are we to believe that these adaptations for cross-fertilization in the Bee Ophrys are absolutely purposeless, as would certainly be the case if this species has always been and will always be self-fertilized … The whole case is perplexing in an unparalled degree, for we have in the same flower elaborate contrivances for directly opposed objects [12]."

The issue still remains a mystery, for according to biologist Bernd Heinrich, "We do not know why this particular orchid, which must have been bee-pollinated at one time in the past, is no longer attracting bees. Possibly the bee species died out, or the bees have become smarter and avoid falling victim to the flower's trick. Should the flower, at some time in the future, again find it greatly advantageous to be cross-pollinated, it will have to perfect its strategy of duping bees [13]."

This duality is indeed perplexing, as one look at the orchid flower makes evident (Fig. 4.17).

References

1. *Contrivances*, 2nd ed., p.202.
2. L.J. Nielsen and B.L. Møller, Scent emission profiles from Darwin's orchid – *Angraecum sesquipedale*: Investigation of the aldoxime metabolism using clustering analysis, *Phytochem.* 120, 3–18, 2015.
3. D.J. Martins and S.D. Johnson, Hawkmoth pollination of aerangoid orchids in Kenya, with special reference to nectar sugar concentration gradients in the floral spurs, *Amer. J. Bot.*, 94, 650–659, 2007.
4. Alfred Russel Wallace, Creation by Law, *Quart. J. Sci.* 471-488, 1867. This essay is a spirited defense of Darwin

and the Theory of Evolution that is actually a review of the Duke of Argyll's book, *The Reign of Law,* that argues the Creationist position.

5. L.T. Wasserthal, The Pollinators of the Malagasy Star Orchids *Angraecum sesquipedale, A. sororium* and *A. compactum* and the evolution of extremely long spurs by pollinator shift, *Plant Bio.*, 110, 343–359, 1997.

6. C. Micheneau *et al.*, Orthoptera, a new order of pollinator, *Ann. of Bot.* 105: 355–364, 2010; S. Hugel *et al.*, *Glomeremus* species from the Mascarene islands (Orthoptera, Gryllacrididae) with the description of the pollinator of an endemic orchid from the island of Réunion, *Zootaxa*, 2545, 58–68, 2010.

7. J. Jersková, *et al.*, Mechanisms and evolution of deceptive pollination in orchids, *Biol. Rev.,* 219–235, 2006.

8. E. Newman *et al.*, Flower colour adaptation in a mimetic orchid, *Proc. R. Soc. B,* 279, 2309–2313, 2012; S.D. Johnson, Evidence for Batesian mimicry in a butterfly-pollinated orchid, *Biol. J. Linn. Soc.* 53, 91–104, 1994.

9. N.A. Van Der Cingel, *An Atlas of Orchid Pollination: European Orchids*, CRC Press, 2001, p.113.

10. Ke-Wei Liu, Zhong-Jian Liu *et al.*, Pollination: Self-fertilization strategy in an orchid, *Nature* 441, 945–946, 22 June 2006.

11. Li-Jun Chen, Zhong-Jian Liu *et al.*, The Anther Steps onto the Stigma for Self-Fertilization in a Slipper Orchid, *Plos One*, 7(5), pp 6, 2012.

12. *Contrivances*, 2nd ed., p.56/57.

13. Bernd Heinrich, *Bumblebee Economics*, Harvard University Press, 2004, p.200.

The lovely flowers embarrass me,
They make me regret I am not a bee –

Emily Dickinson

Some two-thirds to three-quarters of orchid species give off aromatic fragrances, which can be less demanding for the plant than producing nectar. The scent is produced by special gland cells called *osmophores* that may be located on the labellum, petals, or sepals. Analysis from 150 species of 25 genera of orchids shows the presence of approximately 50 different compounds [1]. On average, an individual orchid can produce seven to ten separate compounds that it composes to suit the pollinator. A common organic compound found in more than half of species sampled was *eucalyptol*, which is commonly used for medicinal purposes, in flavoring, and in cosmetics.

It would be nice if all orchids were scented, but unfortunately, this is not the case. Many of the most enchanting modern *Phalaenopsis* hybrids have no scent. The ability to transfer the scent from a fragrant species to one that is not can be problematic. This is because many scented and scentless species are not cross-compatible, and when there has been successful cross-breeding, the progeny may have a weakened scent or even none at all [2].

Some scents are like an intoxicating perfume, while others are quite pungent, horribly putrid, or simply not detectable by the human nose. Of course, the scents are not there for our benefit but instead exist to attract pollinators. Indeed, orchids emit a whole range of different scents: cinnamon (*Phalaenopsis violacea, Lycaste aromatica, Catasetum maculatum* – arising from the compound methyl cinnamate), coconut (*Maxillaria tenuifolia*), citrus (*Rhynchostylis gigantea, Haraella odorata* (Fig. 5.1)), honey (*Bulbophyllum ambrosia,*

Prosthechea cochleata). Some have adopted the fragrance of other flowers, such as: jasmine (*Aeranthes grandalena, Neofinetia falcata*), gardenia (*Brassavola nodosa,* whose fragrance comes out at night to attract pollinating moths), hyacinths/freesias (*Zygopetalums*), roses (*Miltoniopsis santanaei*), lily of the valley (*Pleurothallis pterophora*). Some *Cattleya* scents smell like a whiff from a bottle of French perfume. And in Chap. 3, we already encountered the wonderfully chocolate-scented *Oncidium* Sharry Baby.

Various perfumes have been inspired by the exotic scents of orchids and even bear orchid names, but it would take too many flowers to actually distill the fragrance from them [3]. We will save some of the putrid-scented orchids for the last section.

Cupid's Arrow

Another remarkable orchid genus studied by Darwin was *Catasetum*, located in much of Central and South America. Darwin found that, unusual for orchids, the plants have either male or female flowers that look very different from one another. Further, the male has evolved a very ingenious device to transfer its pollinia to the female.

The small pollinarium projectile containing the pollen masses is safely tucked away in a protective sheath, leaving two touch-sensitive antennae protruding over the lip of male flowers (Fig. 5.2). While the flower does not offer any nectar reward, the lip does give off a cocktail of volatile aromatic fragrances that a male euglossine bee will be

Fig. 5.1 This miniature orchid from Taiwan, *Gastrochilus retrocallus* (*Haraella odorata*), looks like it might smell like citrus, and it does (Image courtesy Sylvia Kappl)

Fig. 5.3 Male *Catasetum viridiflavum* flowers and a too-inquisitive bee with pollinia attached to its head (Image courtesy Mabelín Santos)

Fig. 5.2 *Catasetum callosum* has two touch-sensitive antennae that hover above the lip (Image courtesy Ian Ek Lim)

highly attracted to. It will then scratch at the lip (with special brushes attached to its forelegs) in order to garner some of the scent, which it attaches to special collectors on its hind legs. In doing so, it

will likely disturb the antennae and trigger the plant into action. The pollinarium is shot out disk-first in order to affix itself to the pollinating bee, as the sticky glue sets quickly (Fig. 5.2).

With high-speed photography, the average speed of this projectile has been measured to be 2 meters per second – faster than the striking head of a Pit Viper [4]. Once the hapless bee has been impacted by the pollinarium now stuck to its back, the trauma is sufficient to put it off another male flower (once bitten, twice shy) and instead choose a female to visit, completing its fertilization mission when the pollen masses break off on the sticky stigma. This little shock tactic is a very clever and effective mechanism in directing the bee pollinator to the opposite sex on its next *Catasetum* visitation, and it displays a very choreographed dance between orchid and pollinator. The collection of scent by the male is thought to be for some purpose regarding the attraction of a female, perhaps for a similar reason that a man slaps on aftershave!

Darwin mentions, "Several persons have told me that, when touching the flowers of this genus in their hothouses, the pollinia have struck their faces." So, of course the great scientist had to see for himself. Holding the plant about a meter away from a window, he triggered the antennae, and sure enough, the pollinarium flew out and stuck to the pane of glass (Fig. 5.3).

Fig. 5.4 Note the striking difference: *Catasetum saccatum,* male resupinate flower (L), and the female non-resupinate flower (R). The two male antennae dangling above the lip are primed for action (Images courtesy Juan Fernández Gómez)

Unlike the hermaphroditic flowers of the majority of orchid genera, most species of the *Catasetum* (and *Cycnoches)* genus have flowers that are either one sex or the other. These can sometimes be on the same inflorescence, on different inflorescences, or on different plants. The flowers are strikingly different in appearance, with the male being the more colorful of the two. The female flowers are generally a yellow-green and non-resupinate. The amounts of light and temperature seem to be determining factors of the flowers' sex (Fig. 5.4).

Sexual Deception

In spite of the fact that insects have brains (not quite like ours, of course), they are consistently outwitted by orchids in nature. Indeed, as we have seen, the non-rewarding species must rely on cunning and deception, and fortunately, their long evolutionary history has allowed for a wide variety of complex mechanisms to deceive pollinators.

There is one genus of terrestrial European orchids, *Ophrys*, that not only look like female bees but smells like them too, collectively producing over 100 volatile chemicals. It has been found that the most important factor here is not the orchid's singular appearance, but rather the scent of the female pheromone [5]. The artificial pheromones produced by orchid flowers are properly

known as kairomones, but let us persist with the general term pheromones. This orchid is an example of *sexual deception,* whereby a male insect is fooled into thinking he is achieving a moment of splendor, when in actual fact he is simply carrying out the wishes of the orchid plant to spread *its* seed. This deceptive behavior is possibly unique to orchids, but it is practiced in at least 18 orchid genera and more than 400 species [6] (Fig. 5.5).

During the attempted copulation with the *Ophrys* flower (known as *pseudocopulation*), the pollen grains are deposited on the head of the insect, where they are perfectly situated for the next recipient of the bee's amorous attentions. Since in general, different species of bee pollinate different species of orchid, the orchid has achieved its aim and the bees are duped into fleeting bliss.

Although Darwin did study species of *Ophrys* orchids, he admits in his book on the fertilization of orchids that, "Long and often as I have watched plants of the Bee Ophrys (*Ophrys apifera*), I have never seen one visited by any insect." Of course, he also knew that this species was self-pollinated by the wind. In a footnote, he quotes another source that mentions an account where an individual, "witnessed attacks made upon the Bee Ophrys by a bee similar to those of the troublesome *Apis muscorum*." But Darwin was none the wiser and concluded, "What this … means I cannot conjecture."

Fig. 5.5 Clockwise from top: A *Eucera* bee caught *in flagrante delicto* with *Ophrys flavomarginata; Oph. cretica; Oph. speculum (vernixia); Oph. scolopax* (Clockwise images courtesy Hannes Paulus, Joost Riksen, Wiel Driessen, Wiel Driessen)

One person who did observe the pollination of other *Ophrys* species by insects and carefully studied them for many years was Frenchman Maurice-Alexandre Pouyanne, an amateur botanist and judge stationed in Algiers. He published his bizarre findings regarding the visual and scent attraction of bees to *Ophrys* in 1916 [7]. These findings were subsequently verified by the British orchid lover and naturalist Masters John Godfery in a series of papers starting in 1925 [8].

Further studies of pseudocopulation were made independently from 1927 onwards in several articles by Australian amateur naturalist Edith Coleman [9]. She emphasized the important role that the exuded scent of the female pheromone played in the sexual attraction of male wasps to the genus *Cryptostylis* (see next section).

The *Ophrys* orchid is a veritable chemical laboratory, and the scent produced is a mixture of alkanes, alkenes, aliphatic alcohols, aldehydes,

Fig. 5.6 The tuberous roots of the sexually deceptive *Ophrys tenthredinifera* ("Sawfly Orchid") can be ground up to make salep (Image courtesy José Luis Santamaría)

ketones, esters, and sesquiterpenes, all combined in just the correct proportions for the perfect female bee aroma. Nature is genius.

In fact, in one species of female scoliid wasp (*Campsoscolia ciliata*), a major component of their pheromone scent is 9-hydroxydecanoic acid, a rare compound that is almost never produced in plants. That, however, has not stopped the flowers of *Ophrys speculum* from producing its own recently discovered 9-hydroxydecanoic acid in order to better deceive the males of the scoliid wasp species. Such is the success of the chemical deception that male scoliid wasps have a significant preference for having sex with the flowers of *Ophrys speculum* than with their own females [10] (Fig. 5.5).

Another species that can do a remarkable bee imitation is *Tolumnia* (syn. *Oncidium*) *henekenii* from Haiti and the Dominican Republic. The bumblebee-like flower is pollinated by male *Centris* bees, who apparently exhibit both pseudo-copulatory behavior in trying to mate with the flower and

Fig. 5.7 (L): *Tolumnia henekenii,* from Haiti and the Dominican Republic, is thought to be pollinated by a male bee that either tries to copulate with it or attacks the flower while defending its territory against other male bees. (R): *Trichoceros antennifer,* from South America, imitates the appearance of *Paragymnomma* flies in order to attract a male pollinator (Images courtesy (L) Norbert Dank/www.flickr.com/photos/nurelias, (R) Naoki Takebayashi)

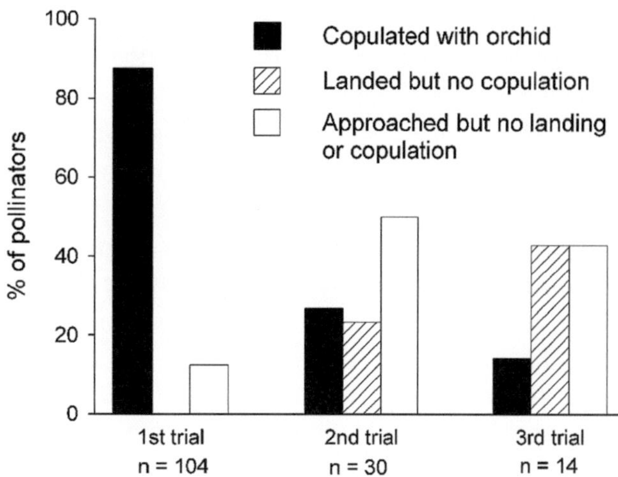

Fig. 5.8 In spite of its name, the male orchid dupe wasp (*Lissopimpla excelsa*) becomes less likely to copulate with *Cryptostylis* orchids with experience. In this experiment, each wasp received three trials, each time being presented with a fresh orchid and its behavior noted (With permission from: Gaskett *et al.* [14])

pseudo-antagonistic behavior by attacking the flower as if were a trespassing male, as we have seen previously with *Oncidium planilabre* (Fig. 5.7).

Not only do orchids imitate bees, but they also mimic flies. Such is the case with the South American *Trichoceros antennifer,* which imitates the appearance of female *Paragymnomma* flies. The sexual attraction seems to be solely based on looks [11].

On the other side of the world, another not very bright male wasp is the "orchid dupe wasp" (*Lissopimpla excelsa)*. Again, from a combination of sight and predominately pheromone scent, the wasp mistakes members of *Cryptostylis* found in parts of Australia for a female wasp, and it will vigorously attempt to copulate with the orchid. Coleman noted that even when the flowers were placed indoors near an open window, the males would come courting. Even though she observed no detectable scent, the male wasps received the pheromone message. The wasps are not fussy and will attempt to copulate with any of the five species members they come across, and frequently even ejaculating on the flower. Of course, the flower is the winner in this charade, succeeding in getting itself pollinated (Fig. 5.8).

This charade is highly successful. According to biologist Amy Martin, "*Cryptostylis* are exceptional deceivers, and also manage to have exceptionally high pollination rates relative to other deceptive orchids – of up to 90% [12]."

Once again, the orchid's sexual deception is so successful that some male pollinators prefer sex with *Cryptostylis* orchids to sex with their own females [13]. This presents an interesting evolutionary situation. On the one hand, wasps who are more discerning and mate with a real female wasp will produce more progeny and should increase in numbers over generations. On the other hand, orchid flowers that produce more exact and alluring copies of the female wasp pheromone are more successful in duping male wasp pollinators and should also produce more progeny over time. This is another example of *coevolution* (Fig. 5.9).

It seems that there is another factor at play here. The female wasps, and those female counterparts of other similarly duped male insects, are almost invariably capable of producing male (only) offspring from unfertilized eggs. Known as *haploids*, they have half the usual number of chromosomes. It is thought that these additional haploid males are an important source in maintaining high levels of pollinators for the orchids they are lured to [14]. The extra males are also seen as a means for the female insects to counterbalance the overpowering allure that the orchids have for their menfolk. The result is a striking example of how devious orchids can interfere with the natural progeny of another species. Interestingly, most sexually deceptive orchids (~90%), have these haploid (as well as diploid) pollinators [15].

Another interesting example of sexual deception can be found in the *Lepanthes* genus, and more specifically in *Lepanthes glicensteinii*. The fool this time is a lustful male fungus gnat [16]. Once again, there is no reward on offer except that of some casual sex. In the study, the authors found that the dark-winged fungus gnat *Bradysia floribunda* was attracted by the pheromone scent of the flower. It would search first on the top side, then on the underside of the disk-shaped leaves, and upon finding a flower, it would assume that it was a female of its own species. The underside of the

Fig. 5.9 (L): The Australian "Large Tongue Orchid," *Cryptostylis subulata,* also established in New Zealand, is pollinated through sexual deception of the male orchid dupe wasp. (R): Not to be outdone is the Australian "Little Tongue Orchid," *Cryptostylis leptochila,* which also makes a dupe of the *Lissopimpla* wasp. Both the Large and Small Tongue orchids were studied by Edith Coleman in her pioneering studies of pseudocopulation (Images courtesy John Varigos (L), Colin and Mischa Rowan (R))

leaf just so happens to be where the fungus gnat would normally expect to find a female fungus gnat waiting for a mate.

The intimate lust between gnat and flower begins with the gnat wrapping itself around the edge of the lip and then swiveling around 180°, with its sexual apparatus remaining perfectly engaged with that of its "female" counterpart. By this maneuver, the pollinarium is attached to the unsuspecting gnat, and the job of the *Lepanthes* is done. Not so the gnat, who, now positioned "tail-to-tail" (as it does with a real female gnat), remains in a lustful embrace that lasts from 3 s to 3 min, with one individual recording a rapturous 20 min thusly engaged. Some evidence suggests that the gnat may actually ejaculate on the flower, and so the deception is complete. Upon making another romantic liaison with another flower of *Lepanthes glicensteinii,* the gnat deposits the pollen mass onto the stigma and helps the orchid achieve its ultimate goal of pollination.

Since other *Lepanthes* species have similar floral structures, it is speculated that a similar mechanism of sexual deception is at play regarding their pollination as well, and this is proving to be the case in further research (Fig. 5.10).

Fig. 5.10 A fungus gnat probing for love with a flower of *Lepanthes glicensteinii* (Image courtesy Mario A. Blanco)

Another deceiver, from Australia again, is the terrestrial *Drakaea,* commonly known as "Hammer Orchids" due to its appearance. The lip is in two parts joined by a hinge, with the outer warty part looking much like a female *thynnid* wasp. The orchid also produces a strong pheromone scent of the female. After the flightless females crawl out of the ground, the orchids will release a strong pheromone scent that can attract numerous suitors, as the females can be a bit sparse (Fig. 5.11).

Fig. 5.11 *Drakaea gracilis*, the slender "Hammer Orchid" of Western Australia (Image courtesy Justin Brown)

One or many suitors will be completely fooled by the *Drakaea* flowers as they hone in on the scent. A successful suitor will attempt to fly the object of its desire off to mate with it. But the orchid has other ideas. Instead, it flips the wasp backward at its midway hinge point and slams the wasp into the pollinia on the opposite side, which sticks to the wasp's thorax. The wasp is then free to fly off and attempt the same lustful act with another *Drakaea*, which leads to pollination.

Australia is full of terrestrial orchid deceivers, another being the remarkable *Caleana major* that looks like a duck, but does not quack like a duck, nor walks like one, but appears to fly like one. It is aptly called the "Flying Duck Orchid." Its scent attracts pollinators such as sawflies, and while they are attempting to mate with the duck's head, the head swings downward, trapping the hapless fly between the head and bowl-shaped body where the column and pollinia are lying in wait. It takes the fly considerable time to free itself, and in the process, it takes the pollinia with it (Fig. 5.12).

Another Australian sexual deceiver is the wondrous *Caladenia dilatata*, ("Green Comb Spider Orchid"), whose flower tips produce a pheromone to attract a male *thynniid* wasp looking for romance. Historically, the root tubers have been consumed for food by the indigenous Aborigines.

Last but not least are members of the genus *Chiloglottis*, which have reproduced on their labellum a perfect replica of a female wasp. According to biologist Anne Gaskett, sexual deception "has

Fig. 5.12 The graceful *Caleana major*, known as the "Flying Duck Orchid" (Image courtesy Malcolm Wells)

evolved at least 6 times in Australia, resulting in 11 sexually deceptive genera there [17] (Fig. 5.13)."

A list of sexually deceptive orchids as of 2011 can be found in Appendix IV. Since then, more have been found, and the author has been informed by Anne Gaskett that an updated list will be available in February 2019. Note the large number of Australian deceivers, largely dominated by *Caladenia*, *Chiloglottis,* and *Pterostylis,* whereas the European deceivers are almost exclusively *Ophrys*.

The success rate of some sexually deceptive orchids in fooling potential pollinators can be extraordinarily high, in some cases 90% or more, and the results for several orchid species have been compiled in Appendix V.

Botanical Role Reversal

In 1879, an Australian journalist by the name of James Hingsley returned from Indonesia and "told tales of a carnivorous red orchid that engulfed butterflies in its petals and devoured them alive." As it turned out, the "orchid" was an insect that had

Fig. 5.13 (L): The masterfully evolved *Caladenia dilatata*. (R): The sexual deceiver, *Chiloglottis reflexa*, with an inviting "female wasp" on its labellum that is already attracting interest (Images courtesy Malcolm Wells)

evolved to mimic the appearance of an orchid. The possibility that an insect could evolve to mimic a flower was already raised by Alfred Russell Wallace in 1877. This has only recently (2014) been scientifically investigated [18].

The insect in question is a praying mantis from Southeast Asia, whose limbs have flaps that resemble orchid petals and sepals. It should be noted, however, that the mantis has evolved a generic flower appearance, but the results are a pretty convincing approximation to an orchid. Common names include "walking flower mantis" and "orchid mantis," and it is the female, which is significantly larger than the male (known as *sexual dimorphism*), who performs the imitation game. Several genera and species of praying mantises are involved in the deception, and their colors can vary from pink, to yellow, to white and banded, although some will change shades slightly to suit the particular required background. It has even been suggested that the floral mimicry could provide camouflage for the praying mantis from its own predators (Fig. 5.14).

So what is the praying mantis up to? It is waiting for insect pollinators to come along, in which case it drops its simulation tactic and pounces on

Fig. 5.14 A Malaysian orchid mantis has evolved to imitate an orchid flower (Image courtesy Chien C. Lee)

the unsuspecting victim for a tasty meal. The much smaller males of the species rely on the traditional methods of stealth and ambush to catch their prey. It is thought that the larger size of the females has to do with their association with the orchid flowers in their insect hunting. The more the mantis evolved to look like an orchid flower (or other flowers), the more it attracted pollinators for food, and increasing their size meant a greater range of insects could be savored, including butterflies, bees, and flies [19].

Moreover, a juvenile female orchid mantis produces the same two compounds, 3HOA and 10HDA, produced by the flowers of *Cymbidium floribundum,* a pheromone mimicry effect known to attract the oriental honeybee. This gives the juvenile orchid mantis a visual and chemical double whammy effect on its prey [20].

Alarming Scents

As we have seen, many non-nectar rewarding orchids use the scent and morphology of a female to attract a male pollinator – a very direct approach to the problem of pollination. But a much more ingenious means has been developed by the Chinese orchid *Dendrobium sinense*. This sweet little thing sends out a chemical scent that is the same as the alarm pheromone given off by two species of honeybees. A hornet (*Vespa bicolor*) is able to detect this chemical scent and assumes it is coming from distressed bees, which it uses to feed its larvae. The hornet actually "pounces" on the source of the pheromone as if attacking its prey, thereby completing the ruse and providing the orchid with a pollinator [21]. What a remarkable evolutionary adaptation! (Fig. 5.15).

Fig. 5.15 Wet and innocent looking, the very clever *Dendrobium sinense* emits a bee alarm pheromone, which attracts a hornet that preys on the bees (Image courtesy Brigitte Meyer)

No Time to Waste

While orchid flowers can remain open for months at a time, some are gone in days, such as those of the Latin American genus *Coryanthes,* known as "Bucket Orchids." These are complexly constructed flowers designed to deceive an unwitting pollinator. They also produce a very pungent smell from a bulbous part of the labellum (the *hypochile*), known to all growers of these unusual orchids. This scent is very attractive to euglossine bees who gather it up, much as is the case for the *Catasetum.* But rather than firing a pollen-laden arrow at the bee, the *Coryanthes* has evolved a much more elaborate mechanism to insure pollination.

As the bee scratches at the surface of the hypochile, often in conflict with other bees, it risks falling into the bucket-shaped lower lip (the *epichile,*), which contains an oily substance secreted by the flower. Covered in this wet substance, the bee has no possibility of flying to freedom. This fact was noted by Darwin, who observed that without the fluid in the lip, the insect would be able to simply fly off without pollinating the flower. There is only a single narrow passageway for the insect to make its escape, aided by small steps. The passageway narrows, the bee struggles to escape, and in doing so, the pollinia attach to its thorax. As the glue quickly sets, the bee eventually finds its release. A similar scenario occurs when the bee enters another *Coryanthes* flower, and the pollinia attach to the sticky stigma, completing the pollination process (Fig. 5.16).

Naturally, the *Coryanthes* orchids do not want a potential bee pollinator to be either too big (it drowns), or too small (it escapes without performing pollination), so the scent of each species is tuned to attract the species of bee of just the appropriate size. This also is an effective deterrent against interbreeding with other *Coryanthes* species in the same vicinity.

A somewhat similar mechanism pertains to the pollination of an allied genus *Stanhopea,* whose species are equally elaborate in their construction and whose flowers generally last 3 days or so. Unusually, the flowers also appear on inflorescences that protrude downward. This is fine in the

Fig. 5.16 Flower of a *Coryanthes speciosa*. Note the two white glands that secret the oily liquid into the lower lip, all a part of the complex mechanism of the pollination process (Image courtesy Dalton Holland Baptista)

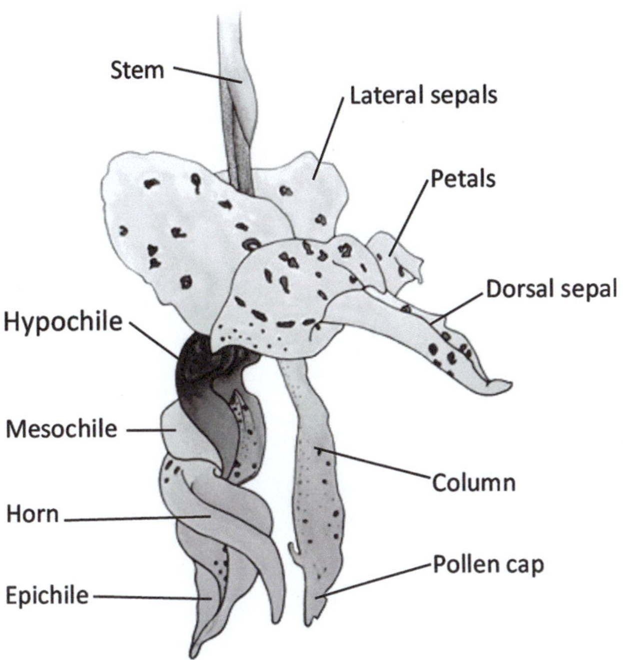

Fig. 5.17 The rather complex architecture of a *Stanhopea* flower. The length of the horns varies in different species (Illustration by Katy Metcalf)

wild when the plant is perched on a tree branch, but in cultivation, it needs to be grown in a slatted basket. The orchid was named for Philip Henry Stanhope, the 4th Earl of Stanhope, who was the president of the Medico-Botanical Society of London from 1829–1837. Begun in 1821 and ceasing its activities in 1852, the society, "…Had for its objects the investigation of the medicinal properties of plants, the study of the *materia medica* of all countries, and the making of awards for original research on the subject." Admirable goals indeed.

The flowers of *Stanhopea,* which number 60 species including six natural hybrids, are highly scented, some overpoweringly so. But male euglossine bees love it. They are drawn by the fragrance as they are to *Coryanthes,* and likewise begin to scratch away at the hypochile. The waxy lip is very slippery and in a vertical position so that the bee can easily slip downward guided by "horns" on each side of the lip past the tip of the long arching column, where the pollinarium becomes fixed to the bee's thorax. Some *Stanhopea* species allow for multiple species of euglossine bee pollinators, correlated with their geographical spread from one region to another, whereas other species permit only one bee species, which has an inhibiting effect on their spread in the wild (Figs. 5.17 and 5.18).

Fig. 5.18 The elaborately constructed *Stanhopea oculata,* typical of the genus. The horns of the labellum can be seen stretching across to the column, which guides the pollinator to its tip (Image courtesy André Heid)

The Big Stink

We have saved the worst for last. There are a few species of orchids that exude an overpoweringly ghastly odor, intentionally so. One such species from Sulawesi and Borneo is *Bulbophyllum*

Fig. 5.19 *Bulbophyllum echinolabium* has a beautifully exotic flower but emits a vile stench of rotten meat (Image courtesy Piotr Markiewicz)

Fig. 5.20 A cluster of *Bulbophyllum beccarii* flowers that you might not want to get too close to but that carrion flies love (Image courtesy Pak Sheikh)

echinolabium. The flowers of this orchid look regal and beautiful – who would guess that they have the putrid stench of rotting meat to attract carrion flies for the purposes of pollination! The flowers can be up to 40 cm (16 in) in size and in some cases can sequentially flower on the same inflorescence for up to a year. As this is not the ideal orchid for a small greenhouse, it is little wonder that its popularity among orchid enthusiasts is limited (Fig. 5.19).

Bulbophyllum echinolabium has some vile companions. One is *Bulbophyllum beccarii*, a native from Borneo, which grows along a very thick rhizome that wends its way up tree trunks seeking the light. The tiny egg-shaped pseudobulbs are dwarfed by the very large yellow-green ovate leaves, slightly funnel-shaped, which nestle snug against the tree. This serves the purpose of catching leaf debris that falls down the trunk and decomposes, providing nutrients for the orchid. Many *Cymbidium* orchids also trap debris with their roots for nourishment.

But the really stunning feature of this amazing orchid is its flowers, which appear in pendulous crimson clusters by the hundreds. The beauty of such a sight is only mitigated by the foul stench of what has been variously described as rotting fish or a herd of dead elephants. Take your pick. Of course, carrion flies find such aromas irresistible (Fig. 5.20).

There are various other vile stinkers in the *Bulbophyllum* genus, such as *Bulb. fletcheranium* ("Tongue Orchid") from southern New Guinea and the related species *Bulb. phalaenopsis,* which has leaves similar to its namesake and is even grown by some brave orchid enthusiasts.

Another perverse example of not being able to judge a book by its cover is the gorgeous yet foulsmelling *Bulbophyllum virescens (binnendijkii)* from Java and Borneo, which produces a circular halo of flowers on foot-long stems (see cover).

So as not to leave the reader thinking that all *Bulbophyllum* have a putrid smell, here is the lovely *Bulbophyllum ambrosia* (syn. *Watsonianum*) from southern China and Vietnam, whose scent is said to be like honey (Fig. 5.21).

Fig. 5.21 *Bulbophyllum ambrosia,* said to have the pleasant smell of honey (Image courtesy André De Kesel, Botanic Garden Meise, Meise, Belgium)

Of course, this is not the end of the orchid story – it is merely the beginning. Or, to quote T.S. Eliot:

We shall not cease from exploration, and the end of all our exploring will be to arrive where we started and know the place for the first time.

References

1. H. Calaway, *et al.*, Biologically active compounds in orchid fragrances, *Science*, June, 1969, 1243–1249.
2. Y.-Y. Hsiao, *et al.*, Research on Orchid Biology and Biotechnology, *Plant Cell Physiol.* 52(9): 1467–1486, 2011.
3. https://www.fragrantica.com/notes/Orchid-154.html
4. C.C. Nicholson, *et al.*, Darwin's bee-trap: The kinetics of *Catasetum*, a new world orchid, *Plant Signal Behavior*, 3(1), 19–23, 2008.
5. T.N. Sherratt, The Evolution of Imperfect Mimicry, *Behavioral Ecol*, 13(6), 2002, 821–826.
6. R.J. Waterman, M.I. Bidartondo, Deception above, deception below: linking pollination and mycorrhizal biology of orchids, *J. Exp. Bot.*, 59(5), 1085–1096, 2008.
7. H. Correvon and M-A. Pouyanne, Un Curieux Cas de Mimétisme ches les Ophrydées, *J. de la Soc. Nat. d'Hort. de France*, Ser. 4, no. 17, 1916. A complete account may be found in the article: Oakes Ames, Pollination of orchids through pseudocopulation, *Botanical Museum Leaflets*, V(1), Harvard University, 1937.
8. M.J. Godfery, The fertilisation of Ophrys Speculum, O. Lutea, and O. Fusca', *J. of Bot., British and Foreign*, 63(2), 33–40, 1925, among others.
9. E. Coleman, Pollination of the orchid *Cryptostylis leptochila. Victorian Naturalist*, 44, 20–22, 1927. E. Coleman, Pollination of *Cryptostylis leptochila. Victorian Naturalist*, 44, 333–340, 1928; E. Coleman, Pollination of an Australian orchid by the male ichneumonid *Lissopimpla semipunctata*, Kirby, *Transactions of the Entomological Society of London*, Part II (Dec): 533–539, 1928. Other papers appeared on the topic of pseudocopulation in subsequent years.
10. M. Ayasse, *et al.*, Attraction in a sexually deceptive orchid by means of unconventional chemicals, *Proc. R. Soc. Lond.*, 2003, 517–522.
11. Nelis A. Van Der Cingel, *An Atlas of Orchid Pollination: European Orchids*, CRC Press, 2001, p.107.
12. A.C. Gaskett, Orchid pollination by sexual deception: pollinator perspectives, *Biol. Rev.*, 86, 33–75, 2011. A list of all the known sexually deceptive orchids and their pollinators as of 2011 can be found in this paper. A few more have been discovered since then.
13. F.P. Schiestl, *et al.*, Chemical communication in the sexually deceptive orchid genus *Cryptostylis*, *Bot. J. of the Linnean Soc.*, 144, 199–205, 2004.
14. A.C. Gaskett, C.G. Winnick, M.E. Herberstein, Orchid sexual deceit provokes ejaculation, *Amer. Nat.*, 171, E206-E212, 2008.
15. *ibid.*
16. M.A. Blanco and G. Barboza, Pseudocopulatory Pollination in *Lepanthes* (Orchidaceae: Pleurothallidinae) by fungus gnats, *Ann. of Bot.*, 95, 763–772, 2005.
17. Anne Gaskett, personal communication; A.C. Gaskett, Orchid pollination by sexual deception: pollinator perspectives, *Biol. Rev.*, 86, 33–75, 2011.
18. J.C. O'Hanlon *et al.*, Pollinator Deception in the Orchid Mantis, *Amer. Naturalist*, 183, 126–132, 2014.
19. G.J. Svenson, *et al.*, Selection for predation, not female fecundity, explains sexual size dimorphism in the orchid mantises, *Sci. Rep.*, 6, 2016.
20. T. Mizuno, *et al.*, "Double-Trick" Visual and Chemical Mimicry by the Juvenile Orchid Mantis *Hymenopus coronatus* used in Predation of the Oriental Honeybee *Apis cerana*, *Zoo. Sci.*, 31, 795–801, 2014.
21. J. Brodmann, *et al.*, Orchid Mimics Honey Bee alarm Pheromone in Order to Attract Hornets for Pollination, *Curr. Biol.*, 19, 1368–1372, 2009.

Appendix I

Ethno-medicinal uses of some important orchid species of Kashmir Himalaya (Adapted with permission from: Gowhar A. Shapoo *et al.*, *International J. of Pharmaceutical and Biological Res.* 4(2), 32–40, 2013)

Sp. No.	Plant species	Vernacular name/English name	Applicable Conditions	Part/parts used	Method of preparation	Dosage	Precaution
1	*Cephalanthera longifolia* (L.) Fritch	Lampatter / Sword leaved Helleborine	1.Weakness	Rhizome	20 gms of rhizome are boiled in 500 ml of milk and 20 gm of dates, 50 gm sugar is added.	The mixture is given twice a day	None
2	*Dactylorhiza hatagirea* D. Don (Soo)	Hatapanja/ Himalayan Marsh orchid	1.Stomachic	Tubers	Dried powders of tubers are mixed water and sugar is added.	100-200 ml/ day	Avoidance of fatty food items
			2.Head ache	Tubers	Fresh tubers are crushed and poultice is prepared.	1–2 times / day	Avoidance of cold water
			3.Fracture	Tubers	The fresh tubers are crushed and mixed with turmeric powder to form a paste by adding little water.	Once a day particularly at bed time.	Avoidance of cold water and complete rest.
			4.Cough and cold	Tubers and flowers	The tubers and dried flowers are boiled in water for about 5 min to get the decoction, and then honey is added to it.	2–3 times a day	Avoidance of cold water, fatty food items and pickles.
			5.Vermifuge	Tubers	Fresh Tubers are crushed and mixed with milk then a little sugar is added.	Once a day at bed time	Avoidance of pickles, cold water and fatty food items.
			6.Weakness of nervous system	Tubers	Fresh Tubers are crushed and mixed with milk then a little honey is added.	Twice a day	Avoidance of cold water
			8.Diarrhea	Tubers and flowers	The tubers and dried flowers are boiled in water for about 5 min to get the decoction.	Twice a day	Avoidance of sugary and fatty food items
			9.Wound healing	Leaves & tubers	Fresh leaves and tubers are crushed and poultice is prepared.	1–2 times / day	Avoidance of cold water.

(continued)

Sp. No.	Plant species	Vernacular name/English name	Applicable Conditions	Part/parts used	Method of preparation	Dosage	Precaution
			10. General weakness after delivery	Whole herb	The powder of dried plant is mixed with ghee.	2–3 times a day	Avoidance of pickles, cold water
			11. Aphrodisiac	Tubers	Fresh Tubers are crushed and mixed with milk then a little sugar and almonds are added.	Once a day at bed time	Avoidance of pickles
3	*Epipactis helleborine* (L.) Crantz.	Sabazl poosh-e-panja/Broad Leaved Helleborine	**1.** Boils	Rhizome	50 gms of rhizome powder is mixed with 80 ml of hot water.	Applied externally on boils.	Avoid cold water
			2. Fever	Leaves	50 gms of leaves are boiled in 500 ml of water.	The decoction is given twice a day	Avoid pickles, fatty food items and cold water
			3. Nerve tonic	Rhizome	20 gms of rhizome are boiled in 500 ml of milk and 10 gms of almond, 10 gms of date and 50 gms of sugar is added.	The mixture is given twice a day	None
			4. Aphrodisiac	Tuber	10 gms of tuber powder is mixed in 300 ml of milk.	The mixture is taken at bed time	Avoid pickles
4	*Epipactis royleana* Lindl.	Wazul poosh-e-panja/Red Flowered Helleborine	**1.** Wounds	Rhizome	The powder of dried rhizome is used to cure wounds	The dried powder is applied on affected body parts	Avoid cold water
			2. General weakness	Rhizome	20 gms of rhizome are boiled in 500 ml of milk and 50 gm sugar is added	The mixture is given twice a day	None
			3. Seminal debility	Rhizome	20 gms of rhizome are boiled in 500 ml of milk and 10 gm almonds, 50 gm sugar is added	The mixture is given twice a day	Avoid pickles
			4. Aphrodisiac	Tuber	10 gms of tuber powder is mixed in 300 ml of milk	Taken at bed time	Avoid pickles

(continued)

Sp. No.	Plant species	Vernacular name/English name	Applicable Conditions	Part/parts used	Method of preparation	Dosage	Precaution
5	*Goodyrea repens* (L.) R.Brown	Meend/ Rattlesnake Plantain	**1.** Toothache	Leaves	The leaves are chewed	The chewed leaves are kept under the affected teeth for few minutes	Avoid sugary food items
			2. Wounds	Whole herb	100 gm of herb is crushed in 50 ml of water	The poultice is applied on affected body parts	Avoid cold water
			3. Loss of appetite	Whole herb	100 gms of whole herb is boiled in 300 ml of water	The decoction is given to patient twice a day	Avoid pickles and fatty food items
			4. Urinary irritation	Whole herb	200 gms of whole herb is boiled in 2liters of water	Decoction is given to patient twice a day	Avoid pickles and fatty food items
			5. Irregular menstruation	Roots & leaves	200 gms of roots and leaves are boiled in 300 ml of water	Decoction is given to patient twice a day	Avoid pickles and fatty food items
			6. Insect bites	Roots & leaves	The roots and leaves are crushed	The juice of plants parts is used to give soothing effects of insect bites and scratches	None
6	*Listera ovata* (L.) R. Brown	Chareed/ Twayblade	**1.** Stomach ailments	Whole herb	50 gm of whole herb is boiled in water 300 ml of water.	The decoction is given to patient twice a day	Avoid fatty and sugary food items
			2. Skin diseases	Rhizome	15 gm of flower powder is mixed with 20-30 ml of oil	The paste is applied on affected skin	Avoid cold water
			3. Indigestion	Whole herb	100 gm of whole herb is boiled in water 300 ml of water.	The decoction is given to patient twice a day	Avoid fatty and sugary food items
			4. Tonic	Leaves	500 gm of leaves are cooked as vegetable	The vegetable is used as tonic	None

(continued)

Sp. No.	Plant species	Vernacular name/English name	Applicable Conditions	Part/parts used	Method of preparation	Dosage	Precaution
7	*Spiranthes sinensis* (Pers.) Ames	Masti-loth/ Lady's Tresses	**1.** Skin eruptions	Flowers	The powder of dried flower are mixed with mustard oil	The mixture of dried flower powder and oil is applied on affected body parts	Avoid cold water
			2. Weakness	Tubers	The tubers are fried and taken with meals	Vegetable is taken with meals twice a day	None
			3. Sore throat	Tubers	50 gms of tubers are mixed with 100 ml of milk	The mixture is applied externally around gullet	Avoid pickles, fatty and sugary food items and dust
			4.Cough and cold	Whole herb	50 gms of whole herb is boiled in 300 ml of water	The mixture is taken orally	Avoid pickles, fatty and food items and cold water
			5. Swelling	Tubers	The tubers are crushed and salt is added and applied on affected parts	Crushed tubers and salt is applied externally	Avoid cold water
			6. Wounds	Tubers	The powder of dried tubers is used to cure wounds	The dried powder is applied on affected body parts	Avoid cold water
			7. Fever	Whole herb	100 gms of whole herb is boiled in 300 ml of water	The decoction is given to patient twice a day	Avoid pickles, fatty food items and cold water

Appendix II

(Top) The flower of *Angraecum sesquipedale* dissected into the different floral parts. (Bottom) Release of volatiles from specific floral parts of *Angraecum sesquipedale* listed as a sequence of organic chemical group clusters. The release of each volatile is displayed in an intensity score (Z) for each floral part with a darker color indicating a higher release than average (0) (color key on the right). White indicates that the given compound has not been found. Note that different chemical groups are produced more intensively from different floral parts. (Adapted with permission from: L.J. Nielsen and B.L. Møller, Scent emission profiles from Darwin's orchid – *Angraecum sesquipedale*: Investigation of the aldoxime metabolism using clustering analysis, *Phytochem.* 120, 3–18, 2015)

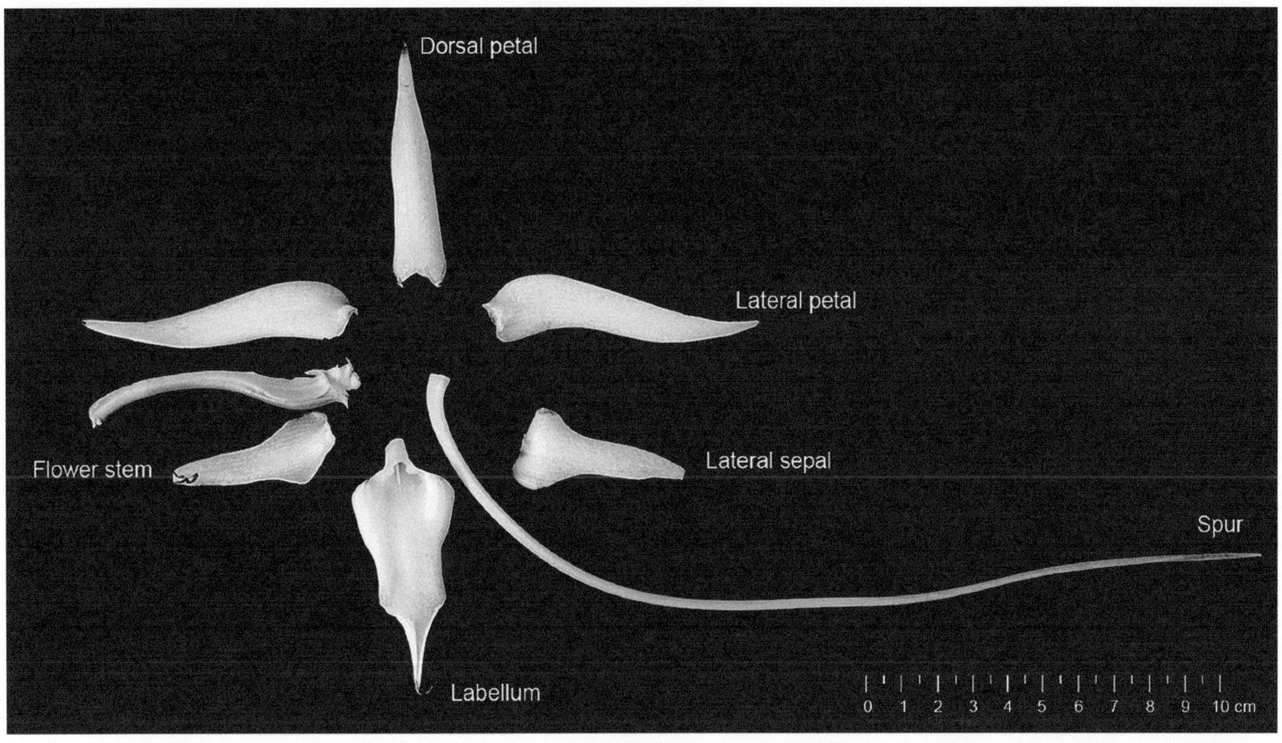

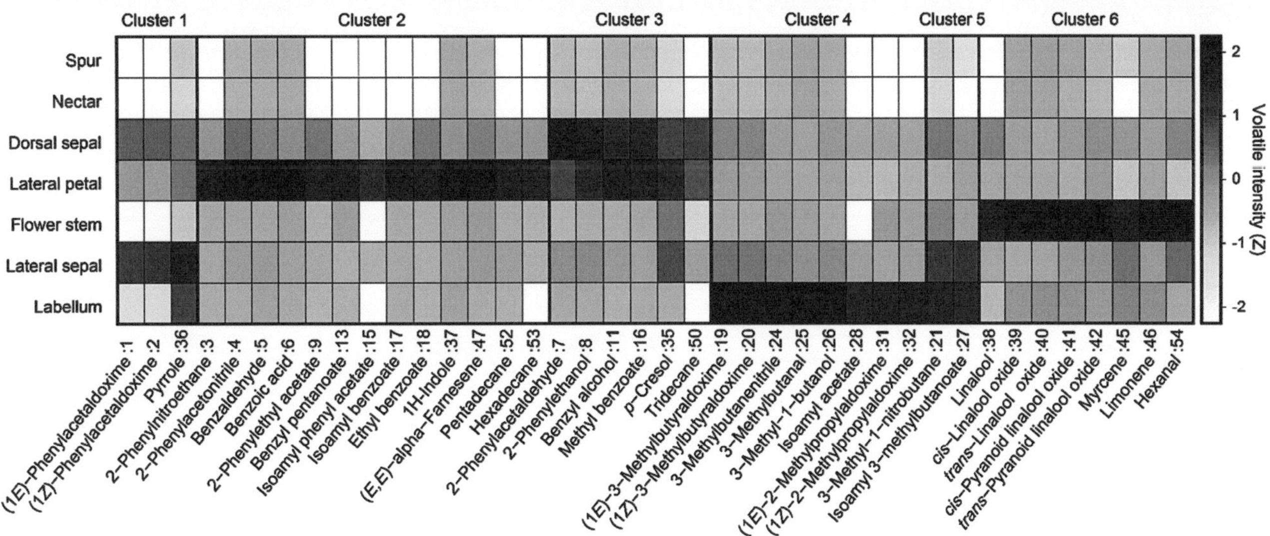

Appendix III

Food-deceptive or unrewarding orchid species tested for similarity in colour with co-occurring plant species that do provide nectar rewards for their pollinators. Pollination success for each orchid species is the average of values reported in the source. N = sum of sample sizes reported in source (Adapted with permission from: Anne C. Gaskett, Orchid pollination by sexual deception: pollinator perspectives, *Biol. Rev.*, 86, 33–75, 2011)

Orchid species	Rewarding model	Pollinator	Orchid mean pollination success
Diseae			
Disa cephalotes Rchb.f. ssp. *cephalotes*	*Scabiosa columbaria* L. (Dipsacaceae)	Long-proboscid flies (Tabanidae, Nemestrinidae)	flowers: 13.9% pollinia removed, 21.2% pollinia deposited (N = 37)
Disa ferruginea (Thunb.) Sw.	*Tritoniopsis triticea* (Burm. F.) Goldbl. (Iridaceae) or *Kniphofia uvaria* (L.) Hook (Asphodelaceae)	Butterfly *Meneris tulbaghia* L. (Nymphalidae)	with model sp.: 69.5% of flowers set fruit (N = 600), without model sp.: 28% (N = 852)
Disa nervosa Lindl.	*Watsonia densiflora s.l.* Baker (Iridaceae)	Long-proboscid fly *Philoliche aethiopica* Thunberg (Tabanidae)	24.9% of flowers pollinated (N = 95); 41.3% fruit set per plant (N = 28)
Disa nivea H.P.Linder	*Zaluzianskya microsiphon* (Kuntze) K. Schum. (Scrophulariaceae)	Long-proboscid fly *Prosoeca ganglbaueri* Lichtwardt (Nemestrinidae)	92.5% of flowers set fruit (N = 890)
Disa pulchra Sond.	*Watsonia lepida* N.E. Brown (Iridaceae)	Long-proboscid fly *Philoliche aethiopica* (Thunberg) (Tabanidae)	15.7% fruit set per plant (N = 17)
Diurideae			
Diuris aequalis Fitzg.	*Gompholobium huegelii* Bentham *(Fabaceae)*		
Diuris maculata Smith	*Daviesia ulicifolia* Andrews ssp. *ulicifolia*, plus other related legumes (Fabaceae)	Male solitary bees *Trichocolletes venustus* (Smith) (Colletidae)	17% of flowers set fruit (N = 122); 59% of plants set fruit (N = 29)
Orchidoidae			
Orchis israelitica H. Baumann & Dafni	*Bellevalia flexuosa* Boiss. (Liliaceae)	Solitary bees *Anthophora* sp., *Eucera clypeata* Erichson (Anthophoridae), *Bombylius* sp. (Bombyliidae)	with model sp. 48.6% of flowers set fruit (N = 692), without model sp. 3.75% (N = 428)
Limodorinae			
Cephalanthera rubra (L.) Rich.	*Campanula* spp. (Campanulaceae)	Solitary bees *Chelostoma fuliginosum* Panzer, *C. campanularum* (Kirby) (Megachilidae)	8.5% fruits per flower (N = 200)

Appendix IV

Sexually deceptive orchid species and their pollinators. All pollinators are male, except where a female pollinator is specifically stated. A question mark (?) after the pollinator name indicates a putative rather than confirmed pollinator. When data are reported from more than one study, symbols (*, †, or ‡) correspond between the data and citation. For Growth form, E = epiphytic, T = terrestrial. For Pollinator sexual behaviour stimulated by the orchid: 1 = pollinator copulates and ejaculates on the orchid, 2 = pollinator copulates with orchid, 3 = pollinator grips or lifts hinged labellum, 4 = pollinator is briefly trapped and transfers pollinia on escape, unconf. = pollination is likely to involve sexual behaviour, but this is unconfirmed (Adapted with permission from: Anne C. Gaskett, Orchid pollination by sexual deception: pollinator perspectives, *Biol. Rev.*, 86, 33–75, 2011)

Orchid	Growth form	Pollinator	Pollinator Family; Order	Pollinator sexual behaviour
Cymbidieae				
Japan				
Cymbidium pumilum Rolfe	E	*Apis cerana japonica* Fabricius	Apidae; Hymenoptera	unconf.
Diseae				
South Africa				
Disa atricapilla (Harv. ex Lindl.)	T	primarily *Podalonia canescens* (Dahlbom)	Sphecidae; Hymenoptera	2
Disa bivalvata (L. f.) T. Durand & Schinz.	T	*Hemipepsis hilaris* Sm., *H. capensis* Fabricius	Pompilidae; Hymenoptera	2
Diurideae				
Australia				
Arthrochilus huntianus (F.Muell.) Blaxell subsp. *huntianus* (prev. *Spiculaea huntiana* (F.Muell.) Schltr.)	T	*Arthrothynnus huntianus* Brown (not *Rhagigaster* sp. as suggested by Rotherham, 1967)	Tiphiidae; Hymenoptera	3
Arthrochilus irritabilis F.Muell.	T	*Arthrothynnus rufiabdominalis* Brown	Tiphiidae; Hymenoptera	3
Arthrochilus latipes D.L.Jones	T	*Arthrothynnus* sp.	Tiphiidae; Hymenoptera	3
Caladenia (syn. *Arachnorchis*) *amnicola* D.L.Jones	T	*Thynnoides senilis* (Erichson)	Tiphiidae; Hymenoptera	3
Caladenia ampla (D.L.Jones) G.N.Backh.	T	*Phymatothynnus* nr. *nitidus* 1	Tiphiidae; Hymenoptera	3
Caladenia (syn. *Arachnorchis*) *arenicola* Hopper & A.P.Br.	T	*Zaspilothynnus nigripes* Guérin	Tiphiidae; Hymenoptera	3
Caladenia (syn. *Arachnorchis*) *atrovespa* D.L.Jones	T	*Thynnoides gracilis* (Westwood)	Tiphiidae; Hymenoptera	3
Caladenia (syn. *Arachnorchis*) *attingens* Hopper & A.P.Br. subsp. *attingens*	T	*Macrothynnus* sp.	Tiphiidae; Hymenoptera	3
Caladenia (syn. *Arachnorchis*) *aurulenta* (D.L.Jones) R.J.Bates	T	Thynnine wasp	Tiphiidae; Hymenoptera	3

(continued)

Orchid	Growth form	Pollinator	Pollinator Family; Order	Pollinator sexual behaviour
Caladenia (syn. *Arachnorchis*) *australis* G.W.Carr	T	*Phymatothynnus* nr. *nitidus* 1	Tiphiidae; Hymenoptera	3
Caladenia (syn. *Drakonorchis*) *barbarossa* Rchb.f.	T	*Thynnoides* sp.* (*Thynnoides bidens* (Saussure) inaccurate, G. Brown pers. comm.)	Tiphiidae; Hymenoptera	3
Caladenia (syn. *Arachnorchis*) *behrii* Schltdl.	T	*Tachynomyia* sp.	Tiphiidae; Hymenoptera	3
Caladenia (syn. *Arachnorchis*) *branwhitei* D.L.Jones	T	*Phymatothynnus monilicornis* (Sm.) complex	Tiphiidae; Hymenoptera	3
Caladenia (syn. *Arachnorchis*) *brownii* Hopper & A.P.Br.	T	*Zaspilothynnus* sp.	Tiphiidae; Hymenoptera	3
Caladenia (syn. *Jonesiopsis*) *cairnsiana* F.Muell.	T	*Thynnoides* sp. *Phymatothynnus nitidus* Sm.?* *Phymatothynnus victor* Turner[†]	Tiphiidae; Hymenoptera	3
Caladenia (syn. *Arachnorchis*) *calcicola* G.W.Carr	T	*Phymatothynnus* nr. *nitidus* 1	Tiphiidae; Hymenoptera	3
Caladenia callitropha D.L.Jones	T	Thynnine wasp	Tiphiidae; Hymenoptera	3
Caladenia (syn. *Arachnorchis*) *caudata* Nicholls	T	Thynnine wasp	Tiphiidae; Hymenoptera	3?
Caladenia (syn. *Arachnorchis*) *clavigera* A.Cunn. Ex Lindl.	T	*Phymatothynnus monilicornis* (Sm.) complex*, *Lophocheilus anilitatus* (Sm.)[†]	Tiphiidae; Hymenoptera	3
Caladenia (syn. *Arachnorchis*) *clavula* D.L.Jones	T	*Lestricothynnus* sp.	Tiphiidae; Hymenoptera	3
Caladenia (syn. *Arachnorchis*) *concinna* (Rupp) D.L.Jones & M.A.Clem.	T	*Aeolothynnus generosus* (Turner)	Tiphiidae; Hymenoptera	3
Caladenia aff. *Concinna*	T	*Neozeleboria* nr. *volatile* (Sm.)	Tiphiidae; Hymenoptera	3
Caladenia (syn. *Arachnorchis*) *concolor* Fitzg.	T	Thynnine wasp	Tiphiidae; Hymenoptera	3
Caladenia (syn. *Arachnorchis*) *conferta* D.L.Jones	T	Thynnine wasp	Tiphiidae; Hymenoptera	3
Caladenia (syn. *Arachnorchis*) *corynephora* A.S.George	T	Thynnine wasp; *Lestricothynnus modestus* (Sm.)*	Tiphiidae; Hymenoptera	3
Caladenia (syn. *Arachnorchis*) *crebra* A.S.George	T	*Campylothynnus assimilis* Sm. *C. flavopictus* (Sm.)	Tiphiidae; Hymenoptera	3

(continued)

Orchid	Growth form	Pollinator	Pollinator Family; Order	Pollinator sexual behaviour
Caladenia (syn. *Arachnorchis*) *cruciformis* D.L.Jones	T	*Phymatothynnus* nr. *nitidus* 1	Tiphiidae; Hymenoptera	3
Caladenia (syn. *Arachnorchis*) *cruscula* Hopper & A.P.Br.	T	Thynnine wasp	Tiphiidae; Hymenoptera	3
Caladenia (syn. *Arachnorchis*) *decora* Hopper & A.P.Br.	T	*Zaspilothynnus nigripes* Guérin	Tiphiidae; Hymenoptera	3
Caladenia (syn. *Arachnorchis*) *dilatata* R.Br.	T	*Thynnoides gracilis* (Westwood), *Thynnoides* spp.	Tiphiidae; Hymenoptera	3
Caladenia (syn. *Arachnorchis*) *discoidea* Lindl.	T	*Phymatothynnus* sp.	Tiphiidae; Hymenoptera	3
Caladenia dilatata var. rhomboidiformis E.Coleman	T	*Ichneumonid* sp.?	Ichneumonidae; Hymenoptera	3?
Caladenia douglasiorum (D.L.Jones) G.N.Backh.	T	*Phymatothynnus* nr. *nitidus* 1	Tiphiidae; Hymenoptera	3
Caladenia (syn. *Jonesiopsis*) *doutchiae* O.H.Sarg.	T	Thynnine wasp*, *Phymatothynnus?* *nitidus*[†]	Tiphiidae; Hymenoptera	3
Caladenia (syn. *Arachnorchis*) *exstans* Hopper & A.P.Br.	T	*Thynnoides* sp.	Tiphiidae; Hymenoptera	3
Caladenia (syn. *Arachnorchis*) *falcata* (Nicholls) M.A.Clem. & Hopper	T	*Zeleboria marginalis* (Westwood)*[†], *Thynnoides* sp.[†], (*Thynnoides bidens** inaccurate; G. Brown pers. comm.)	Tiphiidae; Hymenoptera	3
Caladenia (syn. *Arachnorchis*) *ferruginea* Nicholls	T	Thynnine wasp	Tiphiidae; Hymenoptera	3
Caladenia (syn. *Jonesiopsis*) *filamentosa* R.Br.	T	*Chilothynnus trochanterinus* Brown	Tiphiidae; Hymenoptera	3
Caladenia (syn. *Arachnorchis*) *fitzgeraldii* Rupp	T	Thynnine wasp	Tiphiidae; Hymenoptera	3
Caladenia (syn. *Arachnorchis*) *formosa* G.W.Carr	T	*Phymatothynnus* nr. *pygidialis* 1	Tiphiidae; Hymenoptera	3
Caladenia (syn. *Arachnorchis*) *fragrantissima* D.L.Jones & G.W.Carr	T	*Phymatothynnus* nr. *pygidialis* 2	Tiphiidae; Hymenoptera	3
Caladenia (syn. *Arachnorchis*) *gardneri* Hopper & A.P.Br.	T	Thynnine wasp	Tiphiidae; Hymenoptera	3
Caladenia (syn. *Arachnorchis*) *georgei* Hopper & A.P.Br.	T	Thynnine wasp	Tiphiidae; Hymenoptera	3
Caladenia (syn. *Arachnorchis*) *gladiolata* R.S.Rogers	T	*Zaspilothynnus* sp.	Tiphiidae; Hymenoptera	3

(continued)

Orchid	Growth form	Pollinator	Pollinator Family; Order	Pollinator sexual behaviour
Caladenia (syn. *Arachnorchis*) *hastata* (Nicholls) Rupp	T	*Lophocheilus villosus* Guérin	Tiphiidae; Hymenoptera	3
Caladenia (syn. *Arachnorchis*) aff. *heberleana* Hopper & A.P.Br.	T	Thynnine wasp	Tiphiidae; Hymenoptera	3
Caladenia (syn. *Arachnorchis*) *huegelii* Rchb.f.	T	*Zaspilothynnus* sp.	Tiphiidae; Hymenoptera	3
Caladenia (syn. *Arachnorchis*) *incrassata* Hopper & A.P.Br.	T	*Zaspilothynnus* sp.	Tiphiidae; Hymenoptera	3
Caladenia (syn. *Arachnorchis*) *infundibularis* A.S.George	T	*Zaspilothynnus* sp.	Tiphiidae; Hymenoptera	3
Caladenia (syn. *Arachnorchis*) *insularis* G.W.Carr	T	*Phymatothynnus* nr. *nitidus* 1	Tiphiidae; Hymenoptera	3
Caladenia (syn. *Arachnorchis*) *integra* E.Coleman	T	*Guerinius* sp.? *, thynnine wasp [†]	Tiphiidae; Hymenoptera	3
Caladenia (syn. *Arachnorchis*) *leptoclavia* D.L.Jones	T	*Phymatothynnus monilicornis* (Sm.) complex	Tiphiidae; Hymenoptera	3
Caladenia (syn. *Arachnorchis*) *lobata* Fitzg.	T	Thynnine wasp (*Thynnoides preissi* & *T. bidens* inaccurate; G. Brown pers. comm.)	Tiphiidae; Hymenoptera	3
Caladenia (syn. *Arachnorchis*) *lowanensis* G.W.Carr	T	*Phymatothynnus* nr. *nitidus* 1	Tiphiidae; Hymenoptera	3
Caladenia (syn. *Arachnorchis*) *macrostylis* Fitzg.	T	*Tachynomyia* sp.	Tiphiidae; Hymenoptera	3
Caladenia (syn. *Arachnorchis*) *magniclavata* Nicholls	T	Thynnine wasp	Tiphiidae; Hymenoptera	3
Caladenia montana G.W.Carr	T	*Phymatothynnus* sp.	Tiphiidae; Hymenoptera	3
Caladenia (syn. *Jonesiopsis*) *multiclavia* Rchb.f.	T	*Tachynomia* sp.	Tiphiidae; Hymenoptera	3
Caladenia (syn. *Arachnorchis*) *necrophylla* D.L.Jones	T	*Thynnoides senilis* (Erichson)	Tiphiidae; Hymenoptera	3
Caladenia (syn. *Jonesiopsis*) *pachychila* Hopper & A.P.Br.	T	Thynnine wasp*; *Phymatothynnus*? *nitidus*[†]	Tiphiidae; Hymenoptera	3
Caladenia (syn. *Arachnorchis*) *parva* G.W.Carr	T	*Lophocheilus anilitatus* (Sm.)	Tiphiidae; Hymenoptera	3

(continued)

Orchid	Growth form	Pollinator	Pollinator Family; Order	Pollinator sexual behaviour
Caladenia (syn. *Arachnorchis*) *pectinata* R.S.Rogers	T	*Zeleboria marginalis* (Westwood) (as *Z. marginatus*); *Zaspilothynnus nigripes* Guérin*	Tiphiidae; Hymenoptera	3
Caladenia (syn. *Arachnorchis*) *phaeoclavia* D.L.Jones	T	*Lophocheilus anilitatus* (Sm.)	Tiphiidae; Hymenoptera	3
Caladenia (syn. *Arachnorchis*) *procera* Hopper & A.P.Br	T	*Zaspilothynnus nigripes* Guérin?	Tiphiidae; Hymenoptera	3
Caladenia (syn. *Jonesiopsis*) *pulchra* Hopper & A.P.Br. (prev. *Caladenia flaccida* subsp. *pulchra* Hopper & A.P.Br.)	T	*Aeolothynnus* sp. (as *Asthenothynnus* sp.)	Tiphiidae; Hymenoptera	3
Caladenia (syn. *Arachnorchis*) *radiata* Nicholls	T	Thynnine wasp; *Zaspilothynnus* sp.*, *Catocheilus affinis* (Guérin)*	Tiphiidae; Hymenoptera	3
Caladenia (syn. *Arachnorchis*) *reticulata* Fitzg	T	*Phymatothynnus* nr. *nitidus* 1*, *Phymatothynnus victor*[†]	Tiphiidae; Hymenoptera	3
Caladenia (syn. *Arachnorchis*) *rhomboidiformis* (E.Coleman) M.A.Clem. & Hopper	T	*Zaspilothynnus nigripes* Guérin	Tiphiidae; Hymenoptera	3
Caladenia (syn. *Arachnorchis*) *richardsiorum* D.L.Jones	T	*Phymatothynnus* aff. *pygidialis*	Tiphiidae; Hymenoptera	3
Caladenia (syn. *Arachnorchis*) *rigida* R.S.Rogers	T	*Phymatothynnus* sp.	Tiphiidae; Hymenoptera	3
Caladenia (syn. *Arachnorchis*) *rileyi* D.L.Jones	T	*Thynnoides* new sp. 'R' Brown	Tiphiidae; Hymenoptera	3
Caladenia (syn. *Arachnorchis*) *robinsonii* G.W.Carr	T	*Phymatothynnus* nr. *nitidus* 1	Tiphiidae; Hymenoptera	3
Caladenia (syn. *Jonesiopsis*) *roei* Benth.	T	Thynnine wasp	Tiphiidae; Hymenoptera	3
Caladenia (syn. *Arachnorchis*) *saxatilis* (D.L.Jones) R.J.Bates	T	*Phymatothynnus nitidus*	Tiphiidae; Hymenoptera	3
Caladenia (syn. *Arachnorchis*) *septuosa* D.L.Jones	T	*Thynnoides mesopleuralis* Turner	Tiphiidae; Hymenoptera	3
Caladenia (syn. *Arachnorchis*) *speciosa* Hopper & A.P.Br.	T	Thynnine wasp	Tiphiidae; Hymenoptera	3
Caladenia (syn. *Arachnorchis*) *stellata* D.L.Jones	T	*Phymatothynnus monilicornis* (Sm.)*, *Phymatothynnus monilicornis* complex*, *Phymatothynnus* sp. 14[†]	Tiphiidae; Hymenoptera	3

(continued)

Orchid	Growth form	Pollinator	Pollinator Family; Order	Pollinator sexual behaviour
Caladenia (syn. *Arachnorchis) stricta* R.J.Bates	T	Thynnine wasp	Tiphiidae; Hymenoptera	3
Caladenia (syn. *Arachnorchis) tesselata* D.L.Jones	T	*Phymatothynnus* nr. *nitidus* 1	Tiphiidae; Hymenoptera	3
Caladenia (syn. *Arachnorchis) tensa* G.W.Carr	T	*Thynnoides* aff. *gracilis*	Tiphiidae; Hymenoptera	3
Caladenia (syn. *Arachnorchis) tentaculata* Schltdl.	T	*Thynnoides pugionatus* Guérin (sp. complex), *T. rufithorax* Turner, *Thynnoides gracilis* (Westwood)	Tiphiidae; Hymenoptera	3
Caladenia aff. *tentacula*	T	*Thynnoides* new sp. 'D' Brown	Tiphiidae; Hymenoptera	3
Caladenia (syn. *Arachnorchis) thinicola* Hopper & A.P.Br	T	*Macrothynnus* sp.	Tiphiidae; Hymenoptera	3
Caladenia (syn. *Arachnorchis) toxochila* Tate	T	*Aeolothynnus generosus* (Turner)	Tiphiidae; Hymenoptera	3
Caladenia (syn. *Arachnorchis) uliginosa* A.S.George	T	Thynnine wasp	Tiphiidae; Hymenoptera	3
Caladenia (syn. *Arachnorchis) valida* (Nicholls) M.A.Clem. & D.L.Jones	T	*Phymatothynnus* sp.	Tiphiidae; Hymenoptera	3
Caladenia (syn. *Arachnorchis) woolcockiorum* D.L.Jones	T	Thynnine wasp	Tiphiidae; Hymenoptera	3
Caladenia (syn. *Arachnorchis) verrucosa* G.W.Carr	T	*Zaspilothynnus* sp. nov. 5	Tiphiidae; Hymenoptera	3
Caladenia (syn. *Arachnorchis) villosissima* (G.W.Carr) D.L.Jones & M.A.Clem	T	*Lophocheilus anilitatus* (Sm.)	Tiphiidae; Hymenoptera	3
Caladenia (syn. *Jonesiopsis) wanosa* A.S.George	T	*Phymatothynnus*? *nitidus*	Tiphiidae; Hymenoptera	3
Caladenia (syn. *Arachnorchis) zephyra* (D.L.Jones) R.J.Bates	T	Thynnine wasp	Tiphiidae; Hymenoptera	3
Caleana major R.Br.	T	*Lophyrotoma leachii* (Kirby)*, *Pterygophorus* sp.[†], *L. cyanea* (Leach)[‡]	Pergidae; Hymenoptera	3
Caleana (syn. *Sullivania) minor* R.Br. (prev. *Paracaleana minor* (R.Br.) Blaxell)	T	*Iswaroides armiger* (Turner) (prev. *Thynnoturneria armiger*)	Tiphiidae; Hymenoptera	3
Calochilus caeruleus L.O.Williams (as *Calochilus holtzei* F.Muell.)	T	*Campsomeris* sp.	Scoliidae; Hymenoptera	2

(continued)

Orchid	Growth form	Pollinator	Pollinator Family; Order	Pollinator sexual behaviour
Calochilus campestris R.Br.	T	*Campsomeris tasmaniensis* Saussure	Scoliidae; Hymenoptera	2
Calochilus cupreus R.S.Rogers	T	Scoliid wasp (also self-pollinates)	Scoliidae; Hymenoptera	2
Calochilus pruinosus D.L.Jones	T	Scoliid wasp (also self-pollinates)	Scoliidae; Hymenoptera	2
Calochilus platychilus D.L.Jones	T	Scoliid wasp (also self-pollinates)	Scoliidae; Hymenoptera	2
Chiloglottis anaticeps D.L.Jones	T	*Neozeleboria* n.sp. 33 Brown	Tiphiidae; Hymenoptera	3
Chiloglottis (syn. *Simpliglottis*) *chlorantha* D.L.Jones	T	*Neozeleboria impatiens* Sm. *N. aff. impatiens*	Tiphiidae; Hymenoptera	3
Chiloglottis diphylla R.Br.	T	*Arthrothynnus latus* Brown, *A. angustus* Brown, *Neozeleboria* nr sp. 25(A)*	Tiphiidae; Hymenoptera	3
Chiloglottis (syn. *Myrmechila*) *formicifera* Fitzg.	T	*Neozeleboria* n.sp. 41 Brown	Tiphiidae; Hymenoptera	3
Chiloglottis aff. *formicifera* 1	T	*Neozeleboria* n.sp. 45 Brown	Tiphiidae; Hymenoptera	3
Chiloglottis aff. *formicifera* 2	T	Thynnine wasp	Tiphiidae; Hymenoptera	3
Chiloglottis (syn. *Simpliglottis*) *grammata* G.W.Carr	T	*Eirone leai* Turner	Tiphiidae; Hymenoptera	3
Chiloglottis (syn. *Simpliglottis*) *gunnii* Lindl.	T	*Eirone* sp.	Tiphiidae; Hymenoptera	3
Chiloglottis (syn. *Simpliglottis*) *jeanesii* D.L.Jones	T	*Neozeleboria* nr *impatiens* 2	Tiphiidae; Hymenoptera	3
Chiloglottis palachila D.L.Jones & M.A.Clem.	T	*Chilothynnus palachilus* Brown	Tiphiidae; Hymenoptera	3
Chiloglottis (syn. *Myrmechila*) *platyptera* D.L.Jones	T	*Neozeleboria* n.sp. 40 Brown	Tiphiidae; Hymenoptera	3
Chiloglottis (syn. *Simpliglottis*) *pluricallata* D.L.Jones	T	*Neozeleboria impatiens* Sm. *N. aff. impatiens*	Tiphiidae; Hymenoptera	3
Chiloglottis aff. *pluricallata* (also referred to as *Chiloglottis* 'bifaria' D.L.Jones m.s)	T	*Neozeleboria tabulata* Brown	Tiphiidae; Hymenoptera	3
Chiloglottis aff. *pluricallata* 1	T	*Neozeleboria tabulata* Brown	Tiphiidae; Hymenoptera	3
Chiloglottis aff. *pluricallata* 2	T	*Neozeleboria* nr *monticola 1*	Tiphiidae; Hymenoptera	3
Chiloglottis aff. *pluricallata* 3	T	*Neozeleboria impatiens* Sm.	Tiphiidae; Hymenoptera	3
Chiloglottis reflexa (Labill.) Druce	T	*Neozeleboria* n.sp. 30 Brown	Tiphiidae; Hymenoptera	3
Chiloglottis reflexa sensu stricta (Tas.)	T	Thynnine wasp	Tiphiidae; Hymenoptera	3
Chiloglottis seminuda D.L.Jones	T	*Neozeleboria* n.sp. 29 Brown	Tiphiidae; Hymenoptera	3

(continued)

Orchid	Growth form	Pollinator	Pollinator Family; Order	Pollinator sexual behaviour
Chiloglottis sphrynoides D.L.Jones	T	*Neozeleboria* n.sp. 3 Brown	Tiphiidae; Hymenoptera	3
Chiloglottis sylvestris D.L.Jones & M.A.Clem.	T	*Neozeleboria* n.sp. 50 Brown	Tiphiidae; Hymenoptera	3
Chiloglottis (syn. *Myrmechila*) *trapeziformis* Fitzg.	T	*Neozeleboria cryptoides* Sm., *Zaspilothynnus* sp.‡	Tiphiidae; Hymenoptera	3
Chiloglottis triceratops D.L.Jones	T	*Neozeleboria carinicollis* Turner	Tiphiidae; Hymenoptera	
Chiloglottis trilabra Fitzg.	T	*Neozeleboria proxima* (Turner)	Tiphiidae; Hymenoptera	3
Chiloglottis (syn. *Myrmechila*) *trullata* D.L.Jones	T	Thynnine wasp	Tiphiidae; Hymenoptera	3
Chiloglottis (syn. *Myrmechila*) *truncata* D.L.Jones & M.A.Clem.	T	*Neozeleboria* aff. *ursitatum* Brown, *N.* sp. Spotted, *N.* aff. *cryptoides* 1, *N. aff cryptoides* 2, *N.* sp. (red/black), *N.* sp. (small black)	Tiphiidae; Hymenoptera	3
Chiloglottis (syn. *Simpliglottis*) *turfosa* D.L.Jones	T	*Neozeleboria* nr *monticola* 2	Tiphiidae; Hymenoptera	3
Chiloglottis (syn. *Simpliglottis*) *valida* D.L.Jones	T	*Neozeleboria monticola* Turner, *N. nitidula* (Turner), *N. cryptoides* Sm.	Tiphiidae; Hymenoptera	3
Chiloglottis aff. *valida 1*	T	*Neozeleboria* nr. *impatiens* 1	Tiphiidae; Hymenoptera	3
Chiloglottis aff. *valida 2*	T	*Neozeleboria* nr. *impatiens*, *Neozeleboria* nr. *monticola* 3	Tiphiidae; Hymenoptera	3
Cryptostylis erecta R.Br.	T	*Lissopimpla excelsa* (Costa)	Ichneumonidae; Hymenoptera	1
Cryptostylis hunteriana Nicholls	T	*Lissopimpla excelsa* (Costa)	Ichneumonidae; Hymenoptera	1?
Cryptostylis leptochila Benth.	T	*Lissopimpla excelsa* (Costa) (prev. *L. semipunctata*)	Ichneumonidae; Hymenoptera	1
Cryptostylis ovata R.Br.	T	*Lissopimpla excelsa* (Costa)	Ichneumonidae; Hymenoptera	1
Cryptostylis subulata (Labill.) Rchb.f.	T	*Lissopimpla excelsa* (Costa)	Ichneumonidae; Hymenoptera	1
Drakaea concolor Hopper & A.P.Br.	T	*Zaspilothynnus gilesi* Turner (as *Hemithynnus gilesi*)*, thynnine wasp†	Tiphiidae; Hymenoptera	3
Drakaea confluens Hopper & A.P.Br.	T	Thynnine wasp	Tiphiidae; Hymenoptera	3
Drakaea glyptodon Fitzg.	T	*Zaspilothynnus trilobatus* Turner*, *Z. dilatatus spiculifera* Turner †	Tiphiidae; Hymenoptera	3
Drakaea gracilis Hopper & A.P.Br.	T	*Zaspilothynnus nigripes* Guérin* (*Thynnoides bidens* inaccurate, G. Brown pers. comm.), *Thynnoides elongata*‡	Tiphiidae; Hymenoptera	3
Drakaea livida J.Drumm. (as *Drakaea elastica* Lindl. in Stoutamire, 1979)	T	*Zaspilothynnus nigripes* Guérin	Tiphiidae; Hymenoptera	3

(continued)

Orchid	Growth form	Pollinator	Pollinator Family; Order	Pollinator sexual behaviour
Drakaea livida x *confluens*	T	*Zaspilothynnus nigripes* Guérin, *Z. dilatatus spiculifera* Turner	Tiphiidae; Hymenoptera	3
Drakaea micrantha Hopper & A.P.Br.	T	Thynnine wasp	Tiphiidae; Hymenoptera	3
Drakaea thynniphila A.S.George	T	*Zaspilothynnus* sp.	Tiphiidae; Hymenoptera	3
Leporella fimbriata (Lindl.) A.S.George	T	*Myrmecia urens* Lower	Formicidae; Hymenoptera	2
Oligochaetochilus lepidus D.L.Jones	T	Mycetophilid fly	Mycetophilidae, Diptera	unconf.
Paracaleana hortiorum Hopper & A.P.Br.	T	Thynnine wasp	Tiphiidae, Hymenoptera	3
Paracaleana (syn. *Sullivania*) *nigrita* (J.Drumm. ex Lindl.) Blaxell	T	*Erione* sp.*, *Labium* sp.†	Tiphiidae, Ichneumonidae; Hymenoptera	3
Pterostylis acuminata R.Br.	T	*Culex* sp. female mosquito?	Culicidae; Diptera	unconf.
Pterostylis arenicola M.A.Clem. & J.Stewart (syn. *Oligochaetochilus arenicolus*)	T	Mycetophilid fly	Mycetophilidae, Diptera	unconf.
Pterostylis aspera D.L.Jones & M.A.Clem. (syn. *Diplodium asperum*)	T	Fly	Diptera	unconf.
Pterostylis (syn. *Oligochaetochilus*) *boormanii* Rupp	T	Fly	Diptera	unconf.
Pterostylis cucullata subsp. *sylvicola* D.L.Jones	T	Fly	Diptera	unconf.
Pterostylis curta R.Br.	T	*Mycomya* sp.	Mycetophilidae; Diptera	unconf.
Pterostylis falcata R.S.Rogers	T	*Culex* sp. female mosquito?	Culicidae; Diptera	unconf.
Pterostylis gibbosa R.Br. (syn. *Oligochaetochilus gibbosus*)	T	*Heteropterna* sp.	Mycetophilidae; Diptera	4?
Pterostylis lepida (syn. *Oligochaetochilus lepidus*)	T	Mycetophilid fly	Mycetophilidae, Diptera	unconf.
Pterostylis (syn. *Linguella*) *nana* R.Br.	T	Fly	Diptera	unconf.
Pterostylis nutans R.Br.	T	Fungus gnat	Mycetophilidae; Diptera	4?
Pterostylis psammophila (D.L.Jones) R.J.Bates (syn. *Oligochaetochilus psammophilus*)	T	Mosquito-like fly	Diptera	unconf.
Pterostylis pusilla R.S.Rogers (syn *Oligochaetochilus pusillus* (R.S.Rogers) Szlach.)	T	Fungus gnat	Mycetophilidae; Diptera	4?
Pterostylis (syn. *Diplodium*) *rogersii* E.Coleman	T	Fungus gnat	Mycetophilidae; Diptera	unconf.

(continued)

Orchid	Growth form	Pollinator	Pollinator Family; Order	Pollinator sexual behaviour
Pterostylis rufa R.Br. (syn *Oligochaetochilus rufus*)	T	Fungus gnat*, fly[†]	Diptera	unconf.
Pterostylis sanguinea D.L.Jones & M.A.Clem. (syn. *Urochilus sanguineus* (D.L.Jones & M.A.Clem.) D.L.Jones & M.A.Clem.)	T	Gnat	Diptera	unconf.
Pterostylis (syn. *Ranorchis*) *sargentii* C.R.P.Andrews	T	Fly	Diptera	unconf.
Pterostylis (syn. *Diplodium*) *scabra* Lindl. (as *Pterostylis constricta* O.H.Sarg.)	T	Fly	Diptera	unconf.
Pterostylis (syn. *Urochilus*) *vittata* Lindl.	T	Gnat	Diptera	unconf.
Spiculaea ciliata Lindl.	T	*Thynnoturneria* sp.*, *Iswaroides* sp.[†]	Tiphiidae; Hymenoptera	3
New Zealand				
Cryptostylis subulata (Labill.) Rchb.f.	T	*Lissopimpla excelsa* (Costa)	Ichneumonidae; Hymenoptera	1?
Pterostylis (syn. *Diplodium*) *alobula* (Hatch) L.B.Moore	T	*Zygomyia* sp.	Mycetophilidae; Diptera	4?
Pterostylis australis Hook.f.	T	*Aneura longipalpis* Tonnoir & Edwards?* *Cerotelion* sp.?*, Fungus gnat[†]	Mycetophilidae, Keratophilidae; Diptera	4?
Pterostylis graminea Hook.f.	T	Fungus gnat	Mycetophilidae; Diptera	4?
Pterostylis patens Colenso	T	Fungus gnat?		4?
Pterostylis trullifolia Hook.f. (syn. *Diplodium trullifolium*)	T			unconf.
South America				
Geoblasta pennicillata (Rchb. f.) Hoehne ex Correa	T	*Campsomeris bistrimacula* (Lep.)	Scoliidae; Hymenoptera	2
Epidendreae				
Central America				
Lepanthes glicensteinii Luer	E	*Bradysia floribunda* Mohrig	Sciaridae; Diptera	1?
Lepanthes wendlandii Reichb. f. (E)	E			?
Maxillarieae	E			
South and Central America	E			
Mormolyca ringens (Lindl.)	E	*Nannotrigona testaceicornis* (Lep.), *Scaptotrigona* sp.	Apidae; Hymenoptera	2
Stellilabium sp.	E	Tachinid fly	Tachinidae; Diptera	2
Telipogon sp.	E	Tachinid fly	Tachinidae; Diptera	2
Tolumnia (prev. *Oncidium*) *henekenii* (R.H.Schomb. ex Lindl.)	E	*Centris* sp.?	Anthophoridae; Hymenoptera	unconf.
Trichoceros antennifera (prev. *T. parviflora*) (H. et B). H.B.K.	E	*Paragymnomma* sp.	Tachinidae; Diptera	2

(continued)

Orchid	Growth form	Pollinator	Pollinator Family; Order	Pollinator sexual behaviour
Trigonidium obtusum Lindley	E	*Plebeia droryana* Friese	Meliponinae; Hymenoptera	2 & 4
Orchidoidae				
Europe				
Ophrys aegaea Kalteisen & H.R.Reinhard	T	*Anthophora orientalis* Morawitz	Anthophoridae; Hymenoptera	2
Ophrys aegirtica P.Delforge	T	*Eucera taurica* Morawitz	Anthophoridae; Hymenoptera	2
Ophrys aesculapii Renz	T	*Andrena paucisquama* Noskiewicz	Andrenidae; Hymenoptera	2
Ophrys africana G.Foelsche & W.Foelsche	T	*Andrena flavipes* Panzer	Andrenidae; Hymenoptera	2
Ophrys alasiatica Kreutz, Segers & H.Walraven	T	*Andrena bimaculata* (Kirby)	Andrenidae; Hymenoptera	2
Ophrys algarvensis D.Tyteca, Benito & M. Walravens	T	*Colletes* sp.	Colletidae; Hymenoptera	2
Ophrys annae Devillers-Tersch. & Devillers	T	*Osmia rufa* (L.)	Megachilidae; Hymenoptera	2
Ophrys apifera Huds.	T	*Eucera* sp.*, *Eucera punctulata* Alfken?[†]	Anthophoridae; Hymenoptera	2
Ophrys apiformis Steud.	T	*Eucera barbiventris* Pérez	Anthophoridae; Hymenoptera	2
Ophrys aprilia Devillers & Devillers-Tersch.	T	*Eucera nigrilabris* Lep.	Anthophoridae; Hymenoptera	2
Ophrys apulica (O.Danesch & E.Danesch) O.Danesch & E.Danesch	T	*Eucera* (prev. *Synhalonia*) *rufa* (=*Tetralonia berlandi* Dusmet)	Anthophoridae; Hymenoptera	2
Ophrys arachnitiformis Gren. & M.Philippe	T	*Colletes cunicularius* (L.); *Andrena sabulosa* subsp. *trimmerana, Osmia aurentula* * (*probably = Osmia aurulenta* Panzer)	Andrenidae, Megachilidae, Colletidae; Hymenoptera	2
Ophrys araneola Rchb.	T	*Osmia bicolor* (Schrank)* inaccurate according to Paulus (2000), *Andrena lathyri* Alfken[†], *Andrena combinata* (Christ)[‡]	Megachilidae, Andrenidae; Hymenoptera	2
Ophrys archipelagi Gölz & H.R.Reinhard	T	*Colletes cunicularius* (L.)	Colletidae; Hymenoptera	2
Ophrys argentaria Devillers-Tersch. & Devillers	T	*Andrena fulvata* Stoeckhert	Andrenidae; Hymenoptera	2
Ophrys argolica H.Fleischm. ex Vierh.	T	*Anthophora plagiata* (Illiger)	Anthophoridae; Hymenoptera	2
Ophrys ariadnae H.F.Paulus (as *O. cretica* subsp. *karpathensis* E.Nelson*)	T	*Melecta albifrons* subsp. *albovaria*	Anthophoridae; Hymenoptera	2
Ophrys arnoldii P.Delforge	T	*Andrena nigroaenea* (Kirby)	Andrenidae; Hymenoptera	2
Ophrys atlantica Munby	T	*Chalicodoma parietina* (Geoffr.)	Megachilidae; Hymenoptera	2

(continued)

Orchid	Growth form	Pollinator	Pollinator Family; Order	Pollinator sexual behaviour
Ophrys attica Boiss. & Orph.	T	*Eucera seminuda* Brullé	Anthophoridae; Hymenoptera	2
Ophrys aurelia P.Delforge, Devillers-Tersch. & Devillers	T	*Chalicodoma parietina* (Geoffr.), *C. pyrenaica* (Lep.)	Megachilidae; Hymenoptera	2
Ophrys aveyronensis (J.J.Wood) H.Baumann & Künkele	T	*Andrena hattorfiana* (Fabricius)	Andrenidae; Hymenoptera	2
Ophrys aymoninii (Breistr.) Buttler	T	*Andrena combinata* (Christ)	Andrenidae; Hymenoptera	2
Ophrys balearica P.Delforge	T	*Chalicodoma sicula* (Rossi)	Megachilidae; Hymenoptera	2
Ophrys basilissa A.Alibertis & H.R.Reinhard	T	*Anthophora nigrocincta* Lep.	Anthophoridae; Hymenoptera	2
Ophrys battandieri E.G.Camus	T	*Andrena vetula* Lep.?	Andrenidae; Hymenoptera	2
Ophrys benacensis (Reisigl.) O.Danesch & E.Danesch	T	*Chalicodoma parietina* (Geoffr.)	Megachilidae; Hymenoptera	2
Ophrys bertolonii Moretti	T	*Chalicodoma parietina* (Geoffr.), *C. pyrenaica* (Lep.)	Megachilidae; Hymenoptera	2
Ophrys bertoloniiformis O.Danesch & E.Danesch	T	*Chalicodoma sicula* (Rossi)*, *C. benoisti* Tkalců[†]	Megachilidae; Hymenoptera	2
Ophrys biancae Macch.	T	*Eucera euroa* Tkalců	Anthophoridae; Hymenoptera	2
Ophrys bilunulata Risso	T	*Andrena flavipes* Panzer	Andrenidae; Hymenoptera	2
Ophrys biscutella O.Danesch & E.Danesch	T	*Anthophora retusa* (L.)	Anthophoridae; Hymenoptera	2
Ophrys blitopertha Paulus & Gack	T	*Blitopertha lineolata* (Fischer von Waldheim)	Scarabaeidae; Coleoptera	2
Ophrys bombyliflora Link	T	*Eucera oraniensis* Lep.[†], *E. algira* Brullé[†], *Eucera* sp.*	Anthophoridae; Hymenoptera	2
Ophrys bornmuelleri M.Schulze	T	*Eucera paulusi* Tkalců *, *E. penicillata* Risch [†]	Anthophoridae; Hymenoptera	2
Ophrys bucephala Gölz & H.R.Reinhard	T	*Eucera curvitarsis* Mocz.	Anthophoridae; Hymenoptera	2
Ophrys calocaerina Devillers-Tersch. & Devillers	T	*Andrena labialis* (Kirby)?	Andrenidae; Hymenoptera	2
Ophrys calypsus M.Hirth & H.Spaethe	T	*Eucera dalmatica* Lep.	Anthophoridae; Hymenoptera	2
Ophrys candica Greuter, Matthäs & Risse	T	*Eucera hispana* Lep.*, *E. hispana* Lep. [†]	Anthophoridae; Hymenoptera	2
Ophrys celiensis (O.Danesch & E.Danesch) P.Delforge	T	*Eucera graeca* Radoszkowski	Anthophoridae; Hymenoptera	2
Ophrys cerastes Devillers & Devillers-Tersch. (as *Ophrys cornuta* (small)?*)	T	*Eucera puncticollis* Mor.?	Anthophoridae; Hymenoptera	2
Ophrys ceto P. Devillers, Devillers-Tersch. & P. Delforge	T	*Eucera euroa* Tkalců?, *E. plumigera*?	Anthophoridae; Hymenoptera	2

(continued)

Orchid	Growth form	Pollinator	Pollinator Family; Order	Pollinator sexual behaviour
Ophrys chestermanii (J.J. Wood) Gölz & H.R.Reinhard	T	*Bombus vestalis* (Geoffr.)	Apidae; Hymenoptera	2
Ophrys cilentana Devillers-Tersch. & Devillers	T	*Andrena florentina* Magretti	Andrenidae; Hymenoptera	2
Ophrys cilicica Schltr.	T	*Argogorytes* sp.	Sphecidae; Hymenoptera	2
Ophrys cinereophila Paulus & Gack	T	*Andrena cinereophila* War.	Anthophoridae; Hymenoptera	2
Ophrys cornutula Paulus	T	*Eucera punctulata* Alfken, *E. signifera*	Anthophoridae; Hymenoptera	2
Ophrys crabronifera Sebast. & Mauri	T	*Anthophora plumipes* (Pallas)	Anthophoridae; Hymenoptera	2
Ophrys creberrima Paulus (as *Ophrys fusca* (small)?*)	T	*Andrena creberrima* Pérez*†, *A. flavipes* Panzer*	Andrenidae; Hymenoptera	2
Ophrys cressa Paulus	T	*Andrena merula* War.?	Andrenidae; Hymenoptera	2
Ophrys cretensis (Baumann & Künkele) Paulus	T	*Andrena vachali* subsp. *creticola*	Andrenidae; Hymenoptera	2
Ophrys cretica (Vierh.) E. Nelson	T	*Melecta tuberculata* Lieftinck	Anthophoridae; Hymenoptera	2
Ophrys delphinensis O.Danesch & E.Danesch (pro hybr.)	T	*Anthophora plagiata* (Illiger)	Anthophoridae; Hymenoptera	2
Ophrys discors Bianca (syn. *O. todaris*)	T	*Eucera euroa* Tkalců	Anthophoridae; Hymenoptera	2
Ophrys drumana P.Delforge	T	*Chalicodoma albonotata* (Radoszkowski)	Megachilidae; Hymenoptera	2
Ophrys dyris Maire	T	*Anthophora atroalba* Lep.	Anthophoridae; Hymenoptera	2
Ophrys elatior Gumpr. ex. Paulus	T	*Tetralonia salicariae* (Lep.)	Apidae; Hymenoptera	2
Ophrys elegans (Renz) H. Baumann & Künkele	T	*Anthophora erschowi* Fedtschenko	Anthophoridae; Hymenoptera	2
Ophrys eleonorae Devillers-Tersch. & Devillers	T	*Andrena morio* Brullé	Andrenidae; Hymenoptera	2
Ophrys episcopalis Poir (as *O. episcopalis* (*maxima*)*)	T	*Eucera* (prev. *Synhalonia*) *rufa* (=*Tetralonia berlandi* Dusmet)*, inaccurate according to (Delforge, 2005)	Apidae; Hymenoptera	2
Ophrys exaltata ten.	T	*Colletes cunicularius* (L.)	Colletidae; Hymenoptera	2
Ophrys explanata (Lojac.) P. Delforge	T	*Chalicodoma sicula* (Rossi)	Megachilidae; Hymenoptera	2
Ophrys fabrella Paulus & Ayasse ex P. Delforge	T	*Andrena fabrella* Pérez	Andrenidae; Hymenoptera	2
Ophrys ferrum-equinum Desf.	T	*Chalicodoma parietina* (Geoffr.)	Megachilidae; Hymenoptera	2
Ophrys flavicans Vis.	T	*Chalicodoma manicata* (Giraud)	Megachilidae; Hymenoptera	2

(continued)

Orchid	Growth form	Pollinator	Pollinator Family; Order	Pollinator sexual behaviour
Ophrys flavomarginata (Renz) H. Baumann & Künkele	T	*Eucera dimidiata* Brullé	Anthophoridae; Hymenoptera	2
Ophrys fleischmannii Hayek	T	*Anthophora sicheli* Radoszkowski	Anthophoridae; Hymenoptera	2
Ophrys fuciflora (F.W.Schmidt) Moench	T	*Eucera nigrescens* Pérez*, *E. longicornis* (L.)*, *Microdon latifrons* Loew*, *M. mutabilis* (L.)*, *Phyllopertha horticola* (L.)[†]	Anthophoridae; Hymenoptera Scarabaeidae; Coleoptera	2
Ophrys fusca Link	T	*Andrena nigroaenea* (Kirby), *Colletes cunicularius* (L.)	Andrenidae, Colletidae; Hymenoptera	2
Ophrys fusca (type II Sicily)	T	*Andrena thoracica* Fabricius, *A. florentina* Magretti	Andrenidae; Hymenoptera	2
Ophrys fusca (type III Sicily)	T	*Andrena sabulosa* subsp. *trimerana*	Andrenidae; Hymenoptera	2
Ophrys cf. *fusca (Ophrys sulcata* Devillers & Devillers-Tersch.*)*	T	*Andrena wilkella* (Kirbyi)	Andrenidae; Hymenoptera	2
Ophrys gackiae P.Delforge	T	*Andrena florentina* Magretti?	Andrenidae; Hymenoptera	2
Ophrys garganica E.Nelson ex O.Danesch & E.Danesch	T	*Andrena carbonaria* (L.)	Andrenidae; Hymenoptera	2
Ophrys gazella Devillers-Tersch. & Devillers	T	*Andrena flavipes* Panzer	Andrenidae; Hymenoptera	2
Ophrys gottfriediana Renz	T	*Chalicodoma* sp.	Megachilidae; Hymenoptera	2
Ophrys gracilis (Büel &Danesch) P.Englmaier	T	*Eucera clypeata* Erichson?	Anthophoridae; Hymenoptera	2
Ophrys grammica (B.Willing & E.Willing) Devillers-Tersch. & Devillers (as *O. herae**, according to Paulus (2006)	T	*Andrena nigroaena* (Kirby)	Andrenidae; Hymenoptera	2
Ophrys grandiflora Ten.	T	*Eucera algira* Brullé?	Anthophoridae; Hymenoptera	2
Ophrys grigoriana G.Kretzschmar & H. Kretzschmar	T	*Xylocopa violacea* (L.)	Anthophoridae; Hymenoptera	2
Ophrys hebes (Kalopissis) E.Willing & B.Willing	T	*Andrena symphiti* Schmiedeknecht	Andrenidae; Hymenoptera	2
Ophrys heldreichii Schltr.	T	*Eucera* (prev. *Synhalonia*) *rufa* (=*Tetralonia berlandi* Dusmet)*[†], *Tetralonia alternans (*Brullé*)*[†]	Apidae; Hymenoptera	2
Ophrys helenae Renz	T	*Eucera longicornis* (L.)?	Anthophoridae; Hymenoptera	2
Ophrys helios Kreutz	T	*Eucera* (*Synhalonia*) *cressa*?	Anthophoridae; Hymenoptera	2
Ophrys herae M.Hirth & H. Spaeth	T	*Andrena thoracica* (Fabricius)	Andrenidae; Hymenoptera	2

(continued)

Orchid	Growth form	Pollinator	Pollinator Family; Order	Pollinator sexual behaviour
Ophrys heterochila (Renz & Taubenheim) P.Delforge	T	*Eucera cypria* Alfken	Anthophoridae; Hymenoptera	2
Ophrys holoserica (N.L.Burm.)		*Eucera &Tetralonia* sp.*, *Eucera clypeata* Erichson[†], *E. longicornis* (L.)[†], and rarely *Phyllopertha horticola* (L.)[‡] & *Microdon* sp.[‡]	Anthophoridae; Hymenoptera, Scarabaeidae; Coleoptera Syrphidae; Diptera	
Ophrys incubacea Bianca ex. Tod.	T	*Andrena morio* Brullé	Andrenidae; Hymenoptera	2
Ophrys insectifera L.	T	*Argogorytes fargeii* (Shuckard), *A. mystaceus* (L.), *Anthobium minutum* F. (Staphylinidae)[†]	Sphecidae; Hymenoptera, Staphylinidae; Coleoptera	2
Ophrys integra (Moggr. & Rchb.f.) Paulus & Gack	T	*Colletes cunicularius* (L.)	Colletidae; Hymenoptera	2
Ophrys iricolor Desf.	T	*Andrena morio* Brullé	Andrenidae; Hymenoptera	2
Ophrys israelitica H.Baumann & Künkele	T	*Andrena flavipes* Panzer	Andrenidae; Hymenoptera	2
Ophrys kotschyi H.Fleischm. & Soó	T	*Melecta tuberculata* Lieftinck	Anthophoridae; Hymenoptera	2
Ophrys lacaitae Lojac.	T	*Eucera eucnemidea* Dours	Anthophoridae; Hymenoptera	2
Ophrys laurensis Geniez & Melki	T	*Andrena schulzi* Strand	Andrenidae; Hymenoptera	2
Ophrys lesbis Gölz & H.R.Reinhard	T	*Andrena curiosa* (Morawitz)	Andrenidae; Hymenoptera	2
Ophrys leucadica Renz (pro hybr.)	T	*Andrena flavipes* Panzer*[†], *A. creberrima* Pérez[†]?	Andrenidae; Hymenoptera	2
Ophrys lojaconoi P.Delforge	T	*Andrena ocreata* (Christ)?	Andrenidae; Hymenoptera	2
Ophrys lucana P.Delforge, Devillers-Tersch. & Devillers	T	*Andrena labialis* (Kirby)	Andrenidae; Hymenoptera	2
Ophrys lucentina P.Delforge	T	*Andrena vulpecula* Kriechbaumer	Andrenidae; Hymenoptera	2
Ophrys lucis (Kalteisen & H.R.Reinhard) Paulus & Gack	T	*Anthophora cf. mucida*	Anthophoridae; Hymenoptera	2
Ophrys lunulata Parl.	T	*Osmia kohlii* Ducke	Megachilidae; Hymenoptera	2
Ophrys lupercalis Devillers-Tersch. & Devillers	T	*Andrena nigroaenea* (Kirby)	Colletidae, Andrenidae; Hymenoptera	2
Ophrys lutea Cav.	T	*Andrena cinerea* Brullé[†], *A. senecionis* Pérez[†], *A. clypella* Strand, *A. hasitata*, *A. nigroolivacea* Dours[†]?, *S. panurgimorpha* Mavromoustakis*, *A. humilis* Imhoff *, *A. humilis* subsp. *prunella**, *A.cinerophila* War.*	Andrenidae; Hymenoptera	2
Ophrys lutea subsp. *melena* Renz)	T	*Andrena transitoria* Morawitz	Andrenidae; Hymenoptera	2

(continued)

Orchid	Growth form	Pollinator	Pollinator Family; Order	Pollinator sexual behaviour
Ophrys lyciensis Paulus, Gügel, D.Rückbr. & U. Rückbr.	T	*Eucera graeca* Radoszkowski	Anthophoridae; Hymenoptera	2
Ophrys mammosa Desf.	T	*Andrena fuscosa* Erichson	Andrenidae; Hymenoptera	2
Ophrys marmorata G.Foelsche & W.Foelsche	T	*Andrena wilkella* (Kirby)	Andrenidae; Hymenoptera	2
Ophrys massiliensis Viglione & Véla	T	*Andrena bicolor* Fabricius	Andrenidae; Hymenoptera	2
Ophrys melitensis (Salk.) Devillers-Tersch. & Devillers	T	*Chalicodoma sicula* (Rossi)	Megachilidae; Hymenoptera	2
Ophrys montis-leonis O.Danesch & E.Danesch (pro hybr.)	T	*Colletes cunicularius* (L.)	Colletidae; Hymenoptera	2
Ophrys morio Paulus & Kreutz	T	*Andrena morio* Brullé	Andrenidae; Hymenoptera	2
Ophrys morisii (Martelli) G.Keller & Soó	T	*Anthophora sicheli* Radoszkowski	Anthophoridae; Hymenoptera	2
Ophrys murbeckii H.Fleischm.	T	*Colletes* sp.	Colletidae; Hymenoptera	2
Ophrys neglecta Parl.	T	*Eucera clypeata* Erichson, *E. oraniensis* Lep.?	Anthophoridae; Hymenoptera	2
Ophrys normanii J.J.Wood (pro hybr.)	T	*Bombus vestalis* (Geoffr.)	Apidae; Hymenoptera	2
Ophrys obaesa Lojac.	T	*Andrena flavipes* Panzer	Andrenidae; Hymenoptera	2
Ophrys omegaifera H.Fleischm. (as *O. omegifera* subsp. *omegaifera* in Paulus & Gack 1990)	T	*Anthophora atroalba* subsp. *agamoides, A. nigriceps* Mor.	Anthophoridae; Hymenoptera	2
Ophrys omegaifera subsp. *dyris* (Maire) Del Prete	T	*Anthophora atroalba* subsp. *atroalba*	Anthophoridae; Hymenoptera	2
Ophrys ortuabis M.P.Grasso & Manca	T	*Andrena hypopolia* Smiedeknecht	Andrenidae; Hymenoptera	2
Ophrys oxyrrhynchos Tod.	T	*Eucera graeca* Radoszkowski	Anthophoridae; Hymenoptera	2
Ophrys pallida Raf.	T	*Andrena orbitalis* Morawitz	Andrenidae; Hymenoptera	2
Ophrys panattensis Scrugli, Pessei & Cogoni (pro hybr.)	T	*Osmia rufa* subsp. *rufa*	Megachilidae; Hymenoptera	2
Ophrys panormitana (Tod.) Soó	T	*Andrena thoracica* (Fabricius)*[†], A. florentina* Magretti*, *A. sabulosa* (Scopoli)[†], *A. sabulosa* subsp. *trimerana**	Andrenidae; Hymenoptera	2
Ophrys panormitana var. *praecox* (Corrias) P.Delforge	T	*Andrena thoracica* (Fabricius), *A. nigroaenea* (Kirby)?	Andrenidae; Hymenoptera	2
Ophrys parvimaculata (O.Danesch & E.Danesch) Paulus & Gack	T	*Eucera nigrescens* Pérez	Anthophoridae; Hymenoptera	2

(continued)

Orchid	Growth form	Pollinator	Pollinator Family; Order	Pollinator sexual behaviour
Ophrys parvula Paulus	T	*Andrena tomora* War.	Andrenidae; Hymenoptera	2
Ophrys passionis Sennen	T	*Andrena carbonaria* (L.)	Andrenidae; Hymenoptera	2
Ophrys passionis var. *garganica* (E.Nelson ex. O.Danesch & E. Danesch) P. Delforge	T	*Andrena carbonaria* (L.)	Andrenidae; Hymenoptera	2
Ophrys phryganae Devillers-Tersch. & Devillers	T	*Andrena panurgimorpha* Mavromoustakis, *A. humilis* Imhoff, *A. tadauchii* Gusenleitner, *A. clypella* subsp. *hasiata*?	Andrenidae; Hymenoptera	2
Ophrys picta Link	T	*Eucera barbiventris* Pérez	Anthophoridae; Hymenoptera	2
Ophrys promontorii O.Danesch & E.Danesch	T	*Osmia mustelina* Gerstaecker	Megachilidae; Hymenoptera	2
Ophrys provincialis (Baumann & Künkele) Paulus	T	*Anthophora atriceps* Pérez	Anthophoridae; Hymenoptera	2
Ophrys reinholdii Spruner ex. Boiss.	T	*Melecta* sp.*, *Anthophora obscura*[†], *Eupavlovskia obscura* Friese[‡], *E. funeraria* Sm.? [‡]	Anthophoridae; Hymenoptera	2
Ophrys sabulosa Paulus & Gack ex P.Delforge	T	*Andrena sabulosa* (Scopoli)	Andrenidae; Hymenoptera	2
Ophrys saratoi E.G.Camus (pro hybr.)	T	*Chalicodoma albonotata* (Radoszkoski)	Megachilidae; Hymenoptera	2
Ophrys scolopax Cav.	T	*Eucera longicornis* (L.)*, *Eucera nigrescens* Pérez*, *E. interrupta* Baer*, *E. (Hetereucera) elongatula* Vachal[†]	Anthophoridae; Hymenoptera	2
Ophrys serotina H.Rolli ex H.F. Paulus	T	*Eucera clypeata* Erichson	Anthophoridae; Hymenoptera	2
Ophrys sicula Tineo (as *O. sicula* (minor)*)	T	*Andrena hesperia* Sm.*[†‡], *A. vulpecula* Kriechbaumer*[‡], *A. merula* War.[‡], *A. taraxaci* Giraud[‡], *A. bicolor* Fabricius[‡]?	Andrenidae; Hymenoptera	2
Ophrys sipontensis R.Lorenz & Gembardt	T	*Xylocopa iris* (Christ)	Anthophoridae; Hymenoptera	2
Ophrys sitiaca Paulus, C. Alibertis & A.Alibertis	T	*Andrena nigroaenea* (Kirby)	Andrenidae; Hymenoptera	2
Ophrys speculum Link	T	*Dasyscolia ciliata* (Fabricius) (as *Campsoscolia ciliata*)	Scoliidae; Hymenoptera	2
Ophrys speculum subsp. *orientalis* (Paulus) Paulus & Salkowski	T	*Dasyscolia ciliata* subsp. *araratensis*	Scoliidae; Hymenoptera	2
Ophrys sphegifera Willd.	T	*Eucera notata* Lep.	Anthophoridae; Hymenoptera	2
Ophrys sphegodes Mill.	T	*Andrena nigroaenea* (Kirby)*[†], *A. barbilabris* (Kirby)*, *A. thoracica* (Fabricius)*, *A. cineraria* (L.)*, *A. limata* Eversmann*	Andrenidae; Hymenoptera	2

(continued)

Orchid	Growth form	Pollinator	Pollinator Family; Order	Pollinator sexual behaviour
Ophrys splendida Gölz & H.R.Reinhard	T	*Andrena squalida* Olivier	Andrenidae; Hymenoptera	2
Ophrys spruneri Nyman	T	*Xylocopa iris* (Christ)	Anthophoridae; Hymenoptera	2
Ophrys subinsectifera C.E.Hermos. & J.Sabando	T	*Sterictiphora furcata* (Villers)	Argidae; Hymenoptera	2
Ophrys sulcata Devillers-Tersch. & Devillers	T	*Andrena flavipes* Panzer, *A. ovulata* (Kirby), *A. wilkella* (Kirby)	Andrenidae; Hymenoptera	2
Ophrys tardans O.Danesch & E.Danesch (pro hybr.)	T	*Eucera taurica* Morawitz?	Anthophoridae; Hymenoptera	2
Ophrys tarentina Gölz & H.R.Reinhard	T	*Osmia tricornis* Latreille	Megachilidae; Hymenoptera	2
Ophrys tarquinia P.Delforge	T	*Andrena tibialis* (Kirby)	Andrenidae; Hymenoptera	2
Ophrys tenthredinifera Willd.	T	*Eucera nigrilabris* Lep. *E. dimidiata* Brullé, *E. clypeata* Erichson, *E. algira* Brullé?	Anthophoridae; Hymenoptera	2
Ophrys tetraloniae W.P.Teschner	T	*Tetraloniella fulvescens* (Giraud) (prev. *Tetralonia f.*)*, *Eucera fulvescens* Giraud[†], *Eucera inulae*[†], but reports of *Tetralonia ruficornis* (Fabricius)* are inaccurate[‡].	Apidae; Hymenoptera	2
Ophrys tommasinii Visiani	T	*Andrena pandellei* Pérez	Andrenidae; Hymenoptera	2
Ophrys thriptiensis Paulus	T	*Andrena bicolor* Fabricius?	Andrenidae; Hymenoptera	2
Ophrys transhyrcana Czerniak.	T	*Andrena morio* Brullé*, *A. fuscosa* Erichson[†]?	Andrenidae; Hymenoptera	2
Ophrys umbilicata Desf.	T	*Eucera gaullei* Vachal, *E. galilaea* Tkalců, *E. spatulata* Gribodo, *E. seminuda* Brullé, *Eucera penicillata* Risch[‡]	Anthophoridae; Hymenoptera	2
Ophrys untchjii (M.Schulze) P.Delforge	T	*Eucera clypeata* Erichson	Anthophoridae; Hymenoptera	2
Ophrys urteae H.F.Paulus	T	*Blitopertha nigripennis* Reitter	Scarabaeidae; Coleoptera	2
Ophrys vernixia Brot.	T	*Dasyscolia ciliata* (Fabricius) (as *Campsoscolia ciliata*)	Scoliidae; Hymenoptera	2
Ophrys villosa Desf. (various forms)	T	*Eucera dimidiata* Brullé, *E. bidentata* Pérez, *E. nigrilabris* Lep. *E. rufitarsis* Friese, *E. curvitarsis* Mocsary?	Anthophoridae; Hymenoptera	2
Ophrys zonata Devillers-Tersch. & Devillers	T	*Andrena flavipes* Panzer	Andrenidae; Hymenoptera	2
Orchis galilaea (Bornmüller et Schultze) Schlechter	T	*Lassioglossum marginatum* Brullé (prev. *Halictus* (*Evylaeus*) *marginatus*)	Halictidae; Hymenoptera	2

Appendix V

Percentages of insects fooled into pollinating some Australian sexually deceptive orchids after initial attraction. *N* = visits observed. *Data combined & averaged from more than one study (Adapted with permission from: Anne C. Gaskett, Orchid pollination by sexual deception: pollinator perspectives, *Biol. Rev.*, 86, 33–75, 2011)

Orchid	Pollinator behaviour	% fooled	*N*
Caladenia (syn. *Arachnorchis*) *tentaculata*	Gripping hinged labellum	7.5	287
Chiloglottis diphylla	Gripping hinged labellum	41.7	24
Chiloglottis (syn. *Myrmechila*) *formicifera*	Gripping hinged labellum	10.5	38
Chiloglottis (syn. *Myrmechila*) *platyptera*	Gripping hinged labellum	28.3	53
Chiloglottis (syn. *Simpliglottis*) *pluricallata*	Gripping hinged labellum	3.85	26
Chiloglottis reflexa	Gripping hinged labellum	24.0	48
Chiloglottis seminuda	Gripping hinged labellum	39.0	79
Chiloglottis trilabra	Gripping hinged labellum	23.23*	2897*
Chiloglottis (syn. *Simpliglottis*) *valida*	Gripping hinged labellum	13.0	46
Cryptostylis erecta	Copulation/ejaculation	90.99*	111*
Cryptostylis subulata	Copulation/ejaculation	92.73*	55*
Drakaea glyptodon	Gripping hinged labellum	21.9	618
Leporella fimbriata	Copulation	60.0	55
Spiculaea ciliata	Copulation	44.0	50

Appendix VI

Alphabetical List of Standard Abbreviations for Natural and Hybrid Generic Names

From the *Sander's List of Orchid Hybrids 3 Year Addendum – 2014-2016*, published by the RHS (2017) reproduced with permission

Acw.	= Aberconwayara
Abr.	= Aberrantia
Acp.	= Acampe
Apd.	= Acampodorum
Acy.	= Acampostylis
A.	= Aceras
Ah.	= Acerasherminium
Actg.	= Aceratoglossum
Acba.	= Acinbreea
Acn.	= Acineta
Ain.	= Acinopetala
Aip.	= Aciopea
Akm.	= Ackermania
Aks.	= Ackersteinia
Aco.	= Acoridium
Apa.	= Acrolophia
Aro.	= Acronia
Acro.	= Acropera
Ada	= Ada
Adh.	= Adachilum
Adg.	= Adacidiglossum
Adcm.	= Adacidium
Adgm.	= Adaglossum
Adn.	= Adamantinia
Adm.	= Adamara
Adps.	= Adapasia
Adl.	= Adelopetalum
Adp.	= Adenocalpa
Ade.	= Adenoncos
Ado.	= Adioda
Adog.	= Adoglossum
Anl.	= Adonclioda
Ans.	= Adoncostele
Aen.	= Aenhenrya
Aac.	= Aerachnochilus
Arg.	= Aerangaeris
Argt.	= Aeranganthes
Aergs.	= Aerangis
Aerth.	= Aeranthes
Aescta.	= Aerasconetia
Aed.	= Aeridachnanthe
Aerdns.	= Aeridachnis
Aet.	= Aeridanthe
Aer.	= Aerides
Aersa.	= Aeridisia

(continued)

Aerdts.	= Aeriditis
Aerctm.	= Aeridocentrum
Aerchs.	= Aeridochilus
Aerf.	= Aeridofinetia
Aergm.	= Aeridoglossum
Aegts.	= Aeridoglottis
Aem.	= Aeridolabium
Aerps.	= Aeridopsis
Athe.	= Aeridopsisanthe
Ads.	= Aeridostachya
Aes.	= Aeridostylis
Aerdv.	= Aeridovanda
Aervsa.	= Aeridovanisia
Ards.	= Aeridsonia
Atom.	= Aeristomanda
Aoe.	= Aëroeonia
Aero.	= Aerovanda
Aeh.	= Aetheorhyncha
Agths.	= Agananthes
All.	= Aganella
Agn.	= Aganisia
Agt.	= Aganopeste
Agsp.	= Agasepalum
Agubata	= Agubata
Aitk.	= Aitkenara
Al.	= Alamania
Agwa.	= Alangreatwoodara
Atc.	= Alantuckerara
Aat.	= Alaticaulia
Atg.	= Alatiglossum
Alc.	= Alcockara
Alxra.	= Alexanderara
Alcra.	= Aliceara
Alna.	= Allenara
Aln.	= Allioniara
Alph.	= Alphonsoara
Alv.	= Alvisia
Amal.	= Amalia
Amals.	= Amalias
Amb.	= Amblostoma
Amn.	= Amenopsis
Am.	= Amesangis
Ams.	= Amesara
Ame.	= Amesiella
Aml.	= Amesilabium
Ami.	= Amitostigma
Amo.	= Amoana
Amp.	= Amparoa
Amph.	= Amphiglottis
Amg.	= Ampliglossum
Anb.	= Anabarlia
An.	= Anacamptiplatanthera
Ant.	= Anacamptis
Ana.	= Anacamptorchis
Agz.	= Anagymnorhiza
Amtg.	= Anamantoglossum
Apk.	= Anaphorkis
Acip.	= Ancipitia
Anc.	= Ancistrochilus
Anh.	= Ancistrolanthe
Astp.	= Ancistrophaius
Add.	= Andascodenia
Ande.	= Andersonara
Andre.	= Andreara
Adt.	= Andreettaea
Are.	= Andreettara
Andw.	= Andrewara
Adk.	= Andrewckara
Agl.	= Angellea
Agd.	= Angida
Ayd.	= Anglyda
Angctm.	= Angraecentrum
Agcp.	= Angraeconopsis
Angsts.	= Angraecostylis
Angcm.	= Angraecum
Ancyth.	= Angraecyrtanthes
Angchs.	= Angraeorchis
Angrs.	= Angrangis
Angtla.	= Angranthellea
Angth.	= Angranthes
Angnla.	= Angreoniella
Alr.	= Angularia
Ang.	= Anguloa
Angcst.	= Angulocaste
Ank.	= Anikaara
Akr.	= Ankersmitara
Anct.	= Anoectochilus
Atd.	= Anoectodes
Ano.	= Anoectogoodyera
Anota	= Anota
Ayp.	= Ansecymphyllum
Asg.	= Anselangis
Aslla.	= Ansellia
Asdm.	= Ansidium
Arpt.	= Anteriocamptis
Ahc.	= Anterioherorchis
Atml.	= Anteriomeulenia
Antr.	= Anteriorchis
Atsp.	= Anterioserapias
Anth.	= Anthechostylis

(continued)

Antg.	= Antheglottis	Ascgm.	= Ascoglossum
Anr.	= Antheranthe	Asgts.	= Ascoglottis
Alla.	= Antilla	Ascps.	= Asconopsis
Apr.	= Apoda-prorepentia	Apn.	= Ascoparanthera
Aea.	= Appletonara	Ascns.	= Ascorachnis
Arcp.	= Aracampe	Ald.	= Ascoralda
Ara.	= Arachnadenia	Asc.	= Ascorella
Arach.	= Arachnis	Asnc.	= Ascorenanthochilus
Act.	= Arachnocentron	Arp.	= Ascorhynopsis
Arnc.	= Arachnochilus	Atm.	= Ascostomanda
Arngm.	= Arachnoglossum	Avd.	= Ascovandanthe
Arngl.	= Arachnoglottis	Asvts.	= Ascovandoritis
Aps.	= Arachnopsirea	Atn.	= Ashtonara
Arnps.	= Arachnopsis	Ash.	= Ashworthara
Anch.	= Arachnorchis	Acdp.	= Aspacidopsis
Arnst.	= Arachnostylis	Acid.	= Aspacidostele
Asy.	= Arachnostynopsis	Alm.	= Aspaleomnia
Aad.	= Aranda	Asp.	= Aspasia
Aran.	= Arandanthe	Apo.	= Aspasiopsis
Arnth.	= Aranthera	Aspsm.	= Aspasium
Ael.	= Areldia	Apz.	= Aspezia
Aret.	= Arethusa	Asid.	= Aspioda
Agu.	= Arguellesara	Aspl.	= Asplundara
Agy.	= Argyrorchis	Aspd.	= Aspodonia
Ari.	= Aristotleara	Aspgm.	= Aspoglossum
Ariz.	= Arizara	Apm.	= Aspomesa
Ard.	= Armanda	Aso.	= Aspopsis
Adc.	= Armandacentrum	Asl.	= Aspostele
Art.	= Armocentron	Ath.	= Athertonara
Arl.	= Armochilus	Ato.	= Atopoglossum
Arm.	= Armodachnis	Ast.	= Australia
Amm.	= Armodorum	Au.	= Australorchis
Arh.	= Arthrochilium	Ayb.	= Ayubara
Aru.	= Arthurara	Bka.	= Backhouseara
Arto.	= Artorima	Bak.	= Bakerara
Ar.	= Arundina	Blga.	= Balaguerara
Asr.	= Asarca	Bdwna.	= Baldwinara
Ascdps.	= Ascandopsis	Blk.	= Balenkezia
Ach.	= Aschersonara	Blf.	= Balfourara
Acc.	= Ascocampe	Bln.	= Ballantineara
Ascda.	= Ascocenda	Blm.	= Balmeara
Acch.	= Ascocentrochilus	Bnfd.	= Banfieldara
Acct.	= Ascocentropsis	Btc.	= Baptichilum
Asctm.	= Ascocentrum	Btcm.	= Bapticidium
Ascln.	= Ascocleinetia	Bpt.	= Baptiguezia
Acd.	= Ascocleiserides	Btk.	= Baptikoa
As.	= Ascodenia	Bpd.	= Baptioda
Afd.	= Ascofadanda	Btta.	= Baptirettia
Ascf.	= Ascofinetia	Bpgm.	= Baptistoglossum
Agsta.	= Ascogastisia	Bapt.	= Baptistonia

(continued)

Brgs.	= Barangis
Bnr.	= Baraniara
Bvl.	= Baravolia
Bbra.	= Barbosaara
Bkt.	= Barcatanthe
Bac.	= Barclia
Bard.	= Bardendrum
Bkn.	= Barkeranthe
Bark.	= Barkeria
Bar.	= Barkidendrum
Bkd.	= Barkleyadendrum
Bknts.	= Barkonitis
Bkm.	= Barkorima
Bry.	= Barkronleya
Ba.	= Barlia
Bos.	= Barlorchis
Ban.	= Barnesara
Brmb.	= Barombia
Btt.	= Bartlettara
Btmna.	= Batemannia
Btst.	= Bateostylis
Bmnra.	= Baumannara
Bllra.	= Beallara
Bdra.	= Beardara
Beg.	= Belgeara
Bpr.	= Bennett-Poeara
Bns.	= Bensteinia
Ben.	= Benthamara
Bza.	= Benzingia
Bek.	= Beranekara
Brg.	= Bergmanara
Bkl.	= Berkeleyara
Brln.	= Berlinerara
Bern.	= Bernardara
Ber.	= Bertara
Bet.	= Bettsara
Bhm.	= Bhumipolara
Bhu.	= Bhutanthera
Bic.	= Bicchia
Bfsa.	= Bifranisia
Bif.	= Bifrenaria
Bifdm.	= Bifrenidium
Bifla.	= Bifreniella
Bifrenl.	= Bifrenlaria
Bi.	= Bifrillaria
Bfa.	= Bifrinlaria
Bny.	= Bilneyara
Bilt.	= Biltonara
Bin.	= Binotia
Bid.	= Binotioda
Bish.	= Bishopara
Blkr.	= Blackara
Blp.	= Blepharochilum
Blet.	= Bleteleorchis
Bti.	= Bletia
Blgts.	= Bletiaglottis
Ble.	= Bletilla
Btd.	= Bletundina
Blu.	= Bleuara
Btz.	= Blietzara
Blma.	= Bloomara
Blr.	= Blumeara
Bon.	= Boelanara
Bgd.	= Bogardara
Boh.	= Bohnhofara
Bnf.	= Bohnhoffara
Bkch.	= Bokchoonara
Bby.	= Bolbicymbidium
Bld.	= Bolbidium
Bol.	= Bollea
Blth.	= Bolleanthes
Bca.	= Bolleochondrorhyncha
Bop.	= Bolleoscaphe
Blptm.	= Bollopetalum
Bnt.	= Bonatea
Bonn.	= Bonniera
Bnp.	= Bonplandara
Boo.	= Bootara
Bor.	= Borwickara
Bo.	= Bouletia
Bov.	= Bovornara
Bow.	= Bowringara
Braa.	= Braasiella
Bra.	= Brachtia
Brade.	= Bradeara
Bz.	= Bradriguezia
Bq.	= Bradriquezia
Brd.	= Bradshawara
Bms.	= Bramesa
Bmt.	= Bramiltumnia
Bpc.	= Brapacidium
Bme.	= Braparmesa
Brap.	= Brapasia
Bil.	= Brapilia
Bdt.	= Brasadastele
Bcd.	= Brascidostele
Brt.	= Brasicattleya
Bll.	= Brasilaelia
Bsd.	= Brasilidium
Bch.	= Brasiliorchis

(continued)

Bly.	= Brasilocycnis	Blt.	= Brolaelianthe
Bsh.	= Brasophonia	Brh.	= Brolaephila
Bsl.	= Brasophrolia	Boc.	= Brolarchilis
Bcn.	= Brassacathron	Brm.	= Bromecanthe
Brsa.	= Brassada	Brom.	= Bromheadia
Bsn.	= Brassanthe	Bgn.	= Brongniartara
Btl.	= Brassattlia	Bit.	= Brossitonia
B.	= Brassavola	Bro.	= Broughtonia
Brc.	= Brassavola-cattleya	Bt.	= Broughtopsis
Brsv.	= Brassavolaelia	Bwna.	= Brownara
Brs.	= Brassia	Brum.	= Brummittara
Bssd.	= Brassidiocentrum	Bym.	= Brymerara
Brsdm.	= Brassidium	Byb.	= Bryobium
Bdm.	= Brassidomesa	Byp.	= Bryopinalia
Broda.	= Brassioda	Bcc.	= Buccella
Bssp.	= Brassiopsis	Bck.	= Buckmanara
Bla.	= Brasso-Cattleya-Laelia	Bui.	= Buiara
Bct.	= Brassocatanthe	Bulb.	= Bulbophyllum
Bcl.	= Brassocatlaelia	Bbr.	= Bulborobium
Bsctt.	= Brassocattlaelia	Bul.	= Bullara
Bc.	= Brassocattleya	Bnc.	= Bunochilus
Bss.	= Brassochilum	Bktra.	= Burkhardtara
Brchs.	= Brassochilus	Burk.	= Burkillara
Bdia.	= Brassodiacrium	Bkw.	= Burkinshawara
Bepi.	= Brassoepidendrum	Burr.	= Burrageara
Bpl.	= Brassoepilaelia	Bys.	= Buyssonara
Brsk.	= Brassokeria	Chz.	= Cahuzacara
Bl.	= Brassolaelia	Calda.	= Caladenia
Blc.	= Brassolaeliocattleya	Can.	= Calaeonitis
Bsltc.	= Brassolaeliocattlonitis	Cal.	= Calanthe
Blpr.	= Brassolaeliophila	Cp.	= Calanthidio-preptanthe
Bsy.	= Brassoleya	Ctp.	= Calanthophaius
Bmc.	= Brassomicra	Calsd.	= Calassodia
Bcp.	= Brassoncidopsis	Calt.	= Caletilla
Bis.	= Brassonitis	Call.	= Callostylis
Br.	= Brassonotis	Clc.	= Calnorchis
Bph.	= Brassophila	Clts.	= Caloarethusa
Bsp.	= Brassophranthe	Clchs.	= Calochilus
Bnts.	= Brassophronitis	Cmta.	= Calomitra
Brp.	= Brassopsis	Cpg.	= Calopogon
Bslc.	= Brassosophrolaeliocattleya	Cpt.	= Calopotilla
Bst.	= Brassostele	Cmd.	= Camaridium
Bstna.	= Brassotonia	Cmt.	= Camarotis
Bv.	= Brassovolaelia	Cam.	= Campanulorchis
Brat.	= Bratonia	Cmpba.	= Campbellara
Blg.	= Brevilongium	Cnn.	= Cannaeorchis
Brn.	= Brianara	Cnz.	= Cannazzaroara
Bba.	= Briggs-Buryara	Cpa.	= Cappeara
Brlda.	= Brilliandeara	Carn.	= Carenidium
Btv.	= Broanthevola	Cwt.	= Carlwithnerara

(continued)

Crml.	= Carmichaelara	Cbd.	= Caulbardendrum
Cro.	= Carolara	Cdc.	= Cauldenclia
Clg.	= Carolineleongara	Cun.	= Cauleytonia
Crp.	= Carparomorchis	Clps.	= Cauliopsis
Cptra.	= Carpenterara	Ckr.	= Caulkeria
Car.	= Carrara	Clty.	= Caulocattleya
Ctra.	= Carterara	Cup.	= Caulophila
Csr.	= Casoara	Cut.	= Caulrianitis
Cat.	= Cataleria	Cuv.	= Caulrianvola
Cag.	= Catamangis	Cul.	= Caultonia
Ctmds.	= Catamodes	Cnph.	= Caultoniophila
Ctnchs.	= Catanoches	Cny.	= Cautonleya
Ctsda.	= Catasandra	Ctg.	= Centrogenium
Ctsl.	= Catasellia	Cen.	= Centroglossa
Ctsm.	= Catasetum	Cnl.	= Centropetalum
Ca.	= Catawesia	Ceph.	= Cephalanthera
Ctc.	= Catcattleyella	Cpts.	= Cephalopactis
Ctll.	= Catcaullia	Chp.	= Cephalophrys
Ctyl.	= Catcylaelia	Cphl.	= Cephalorchis
Ctl.	= Catlaelia	Cph.	= Cephalorhiza
Cnc.	= Catminichea	Ce.	= Cepobaculum
Ctr.	= Cattarthrophila	Cpr.	= Ceporillia
Cka.	= Cattkeria	Cr.	= Ceraia
Cttl.	= Cattlaelia	Cer.	= Ceratobium
Cas.	= Cattlassia	Crtn.	= Ceratocentron
C.	= Cattleya	Crgm.	= Ceratograecum
Ctyh.	= Cattleychea	Csl.	= Ceratosiella
Cyy.	= Cattleychytonia	Cdw.	= Chadwickara
Cte.	= Cattleyella	Cba.	= Chamberlainara
Cttrn.	= Cattleyodendron	Cham.	= Chamodenia
Cdm.	= Cattleyodendrum	Chm.	= Chamorchis
Ctps.	= Cattleyopsis	Cng.	= Changara
Ctpga.	= Cattleyopsisgoa	Chap.	= Chapmanara
Ctpsta.	= Cattleyopsistonia	Cae.	= Charlesara
Cva.	= Cattleyovola	Ckp.	= Charlesknappara
Ctna.	= Cattleytonia	Cha.	= Charlesworthara
Ctt.	= Cattlianthe	Chra.	= Charleswortheara
Ctph.	= Cattoniphila	Charli.	= Charlesworthiara
Ctts.	= Cattotes	Charl.	= Charlieara
Cyi.	= Cattychilis	Chau.	= Chaubardia
Cty.	= Catyclia	Chbth.	= Chaubardianthes
Cauc.	= Caucaea	Chbl.	= Chaubardiella
Cet.	= Caucaerettia	Cbz.	= Chaubewiczella
Cus.	= Cauchostele	Cyh.	= Chelychocentrum
Ccd.	= Caucidium	Chd.	= Chelycidium
Cll.	= Caulaelia	Cey.	= Chelyopsis
Clk.	= Caulaeliokeria	Cly.	= Chelyorchis
Clt.	= Caularstedella	Cna.	= Chenara
Cau.	= Caularthron	Chen.	= Chenlanara
Clv.	= Caulavola	Chew.	= Chewara

(continued)

Chctm.	= Chilocentrum	Cleis.	= Cleisostoma
Chil.	= Chiloglottis	Cst.	= Cleisostomopsis
Chsch.	= Chiloschista	Cli.	= Cleisostylanda
Csg.	= Chilosimpliglottis	Clsty.	= Cleisostylis
Chi.	= Chinheongara	Cltha.	= Cleisotheria
Cgv.	= Chlorogavilea	Cvd.	= Cleisovanda
Cdths.	= Chondranthes	Clst.	= Cleistesiopsis
Chdb.	= Chondrobollea	Clm.	= Clomophyllum
Cho.	= Chondropetalum	Cgh.	= Cloughara
Chdrh.	= Chondrorhyncha	Cwr.	= Cloweandra
Cds.	= Chondroscaphe	Clw.	= Clowenoches
Chu.	= Chouara	Cws.	= Clowesetenaea
Csn.	= Chrisanda	Clo.	= Clowesetum
Chrt.	= Chrisanthera	Cl.	= Clowesia
Cne.	= Chrisnetia	Cwl.	= Clowsellia
Cps.	= Chrisnopsis	Coc.	= Coccineorchis
Cdt.	= Christendoritis	Ccc.	= Coccinoglottis
Chnl.	= Christensonella	Ccr.	= Cochardia
Chri.	= Christensonia	Cnths.	= Cochleanthes
Chn.	= Christenstylis	Cccst.	= Cochlecaste
Chtra.	= Christieara	Cclna.	= Cochlenia
Crc.	= Christocentrum	Colta.	= Cochleottia
Cht.	= Chromatotriccum	Ccptm.	= Cochlepetalum
Chry.	= Chrysocycnis	Cos.	= Cochlesepalum
Chnya.	= Chuanyenara	Clsl.	= Cochlesteinella
Chtn.	= Chuatianara	Czl.	= Cochlezella
Chlt.	= Chyletia	Ccz.	= Cochlezia
Chy.	= Chysis	Cch.	= Cochlicidichilum
Chyt.	= Chytroglossa	Cda.	= Cochlioda
Cra.	= Cirrhaea	Cdp.	= Cochliodopsis
Chpa.	= Cirrhopea	Cit.	= Cochlistele
Cirr.	= Cirrhopetalum	Ccp.	= Cochloncopsis
Crphm.	= Cirrhophyllum	Csp.	= Cochloscaphe
Cstx.	= Cischostalix	Ccl.	= Cochlumnia
Cisch.	= Cischweinfia	Coeln.	= Coeleione
Ccw.	= Cischweinidium	Cga.	= Coeloglossgymnadenia
Clka.	= Clarkeara	Co.	= Coeloglossum
Cgy.	= Claudegayara	Coel.	= Coelogyne
Cdh.	= Claudehamiltonara	Cox.	= Cogniauxara
Cdld.	= Claudiasauledaara	Coh.	= Cohniella
Cay.	= Clayara	Cnlr.	= Cohnlophiaris
Ces.	= Cleisanda	Clx.	= Colax
Csct.	= Cleiscocentrum	Cole.	= Coleara
Clclp.	= Cleisocalpa	C-s.	= Collare-stuartense
Clctn.	= Cleisocentron	Col.	= Collierara
Clsd.	= Cleisodes	Colm.	= Colmanara
Clfta.	= Cleisofinetia	Cmm.	= Commersonara
Clnps.	= Cleisonopsis	Comp.	= Comparettia
Clspa.	= Cleisopera	Cmr.	= Comparumnia
Clq.	= Cleisoquetia	Cpz.	= Compelenzia

(continued)

Com.	= Comperia	Cymla.	= Cymbidiella
Ctgs.	= Comptoglossum	Cdl.	= Cymbidilophia
Coy.	= Conattleya	Cdn.	= Cymbidimangis
Conc.	= Conchidium	Cbn.	= Cymbidinaea
Conph.	= Conphronitis	Cym.	= Cymbidium
Cnt.	= Constanciaara	Cbg.	= Cymbiglossum
Const.	= Constantia	Cbl.	= Cymbiliorchis
Cook.	= Cookara	Cbp.	= Cymbipetalum
Cok.	= Cooksonara	Cml.	= Cymbisellia
Cpp.	= Coppensia	Cbgl.	= Cymboglossum
cppt.	= Coppensitonia	Cymph.	= Cymphiella
Crb.	= Corbettara	Cych.	= Cynorchis
Coa.	= Corningara	Cyk.	= Cynorkaria
Crd.	= Coronadoara	Cyn.	= Cynorkis
Corl.	= Correllara	Ccm.	= Cypercymbidium
Corr.	= Correvonia	Cyd.	= Cyperocymbidium
Crths.	= Coryanthes	Cphd.	= Cyphiopedilum
Cnd.	= Corydandra	Cgd.	= Cyphragmipedium
Crhpa.	= Coryhopea	Cyp.	= Cypripedium
Cyt.	= Corysanthes	Crs.	= Cyrassostele
Cot.	= Cottonia	Cyrtl.	= Cyrtellia
Cow.	= Cowperara	Cto.	= Cyrtidiorchis
Crg.	= Crangonorchis	Cti.	= Cyrtidium
Crv.	= Cravenara	Cip.	= Cyrtionopsis
Craw.	= Crawshayara	Csd.	= Cyrtobrassidium
Cpd.	= Crepidium	Cybs.	= Cyrtobrassonia
Ctcm.	= Cryptocentrum	Cuc.	= Cyrtocaucaea
Crypt.	= Cryptopus	Crt.	= Cyrtochiloides
Cum.	= Cucumeria	Cyr.	= Cyrtochilum
Cud.	= Cuitlacidium	Cil.	= Cyrtocidistele
Cu.	= Cuitlauzina	Ctd.	= Cyrtocidium
Cid.	= Cuitlioda	Cydn.	= Cyrtodenia
Cdg.	= Cuitliodaglossum	Cdo.	= Cyrtodontioda
Ctn.	= Cuitlumnia	Crdc.	= Cyrtodontocidium
Cca.	= Cyanicula	Code.	= Cyrtodontostele
Cnr.	= Cyanthera	Cgl.	= Cyrtoglossum
Cct.	= Cycatonia	Cgt.	= Cyrtoglottis
Cgn.	= Cycgalenodes	Ctgo.	= Cyrtogomestele
Cyl.	= Cycleria	Cgc.	= Cyrtogramcymbidium
Cld.	= Cyclodes	Cyz.	= Cyrtolauzina
Cycl.	= Cyclopogon	Crl.	= Cyrtolioda
Clos.	= Cyclosia	Clr.	= Cyrtollaria
Cycda.	= Cycnandra	Cgp.	= Cyrtomangophyllum
Cyc.	= Cycnoches	Ctea.	= Cyrtonaea
Cycd.	= Cycnodes	Cop.	= Cyrtoncidopsis
Cnp.	= Cycnophyllum	Cdu.	= Cyrtoncidumnia
Cysl.	= Cycsellia	Crn.	= Cyrtoniopsis
Cyln.	= Cylindrolobus	Cpas.	= Cyrtopasia
Cma.	= Cymaclosetum	Cyrt.	= Cyrtopodium
Cymst.	= Cymasetum	Cyrtcs.	= Cyrtorchis

(continued)

Cye.	= Cyrtostele		Dpthe.	= Diaphananthe
Crz.	= Cyrtozia		Dias.	= Diaschomburgkia
Ctu.	= Cyrtumnia		Dich.	= Dichaea
Cys.	= Cysepedium		Dmts.	= Dichromanthus
Dcz.	= Dacruzara		Dcy.	= Dichromarrhynchos
D.	= Dactylanthera		Dct.	= Dichromoglottis
Dps.	= Dactylocamptis		Dhw.	= Dicksonhowara
Dtd.	= Dactylodenia		Die.	= Dienia
Dam.	= Dactyloglossum		Dign.	= Dignathe
Dtylo.	= Dactylorchis		Dill.	= Dillonara
Dact.	= Dactylorhiza		Dlc.	= Dilochiopsis
Datrr.	= Dactylorrhiza		Dmd.	= Dimerandra
Dty.	= Daiotyla		Dph.	= Dimorphachnis
Dlm.	= Dallemagneara		Dpda.	= Dimorphanda
Dnh.	= Danhatchia		Dii.	= Dineclia
Dar.	= Darwinara		Ddo.	= Diodonopsis
Dyg.	= Dasyglossum		Dvl.	= Diovallia
Dj.	= Davejonesia		Dplch.	= Diplochilus
Dvh.	= Davidhuntara		Dpl.	= Diplodium
Dvd.	= Davidsonara		Dpnps.	= Diplonopsis
Dbr.	= Debarriara		Dld.	= Diplopanda
Dbra.	= Debruyneara		Dpra.	= Diploprora
Dgmra.	= Degarmoara		Dipr.	= Diplorrhiza
Dsla.	= Deiselara		Disa	= Disa
Dek.	= Dekensara		Dtl.	= Disticholiparis
Dvx.	= Delouvrexara		Ds.	= Distichorchis
Dn.	= Dendrobates		Diuris	= Diuris
Den.	= Dendrobium		Dix.	= Dixuanara
Dec.	= Dendrocatanthe		Djn.	= Docjonesia
Dc.	= Dendrocattleya		Doc.	= Dockrillia
Ddc.	= Dendrochilum		Dok.	= Dockrilobium
Denga.	= Dendrogeria		Dkb.	= Dockrobium
Ddlr.	= Dendrolirium		Doda.	= Dodara
Dlax.	= Dendrophylax		Don.	= Doinara
Dni.	= Denisara		Dol.	= Dolichopsis
Dpn.	= Deprinsara		Do.	= Domindendrum
Droa.	= Derosaara		Ddma.	= Domindesmia
Dmt.	= Desmetara		Dml.	= Domingleya
Des.	= Desmotrichum		Dga.	= Domingoa
Dvra.	= Devereuxara		Dly.	= Dominleychile
Dvr.	= Devriesara		Dmtna.	= Domintonia
Dew.	= Dewolfara		Dmya.	= Dominyara
Diab.	= Diabroughtonia		Dmlps.	= Domliopsis
Diaca.	= Diacattleya		Dnt.	= Donaestelaara
Diacm.	= Diacrium		Dclna.	= Doncollinara
Diad.	= Diadenium		Dsg.	= Donsutingara
Dkra.	= Diakeria		Ddps.	= Dorandopsis
Dial.	= Dialaelia		Drd.	= Doredirea
Dialps.	= Dialaeliopsis		Dhta.	= Doreenhuntara
Dpgs.	= Diaphanangis		Dctm.	= Doricentrum

(continued)

Dd.	= Doridium
Drlla.	= Doriella
Dllps.	= Doriellaopsis
Dfta.	= Dorifinetia
Drgm.	= Doriglossum
Dpm.	= Doriopsisium
Drsa.	= Dorisia
Dst.	= Doristylis
Dtps.	= Doritaenopsis
Dor.	= Doritis
Dr.	= Doritopsis
Drm.	= Dormanara
Dtha.	= Dorthera
Doss.	= Dossinia
Dsh.	= Dossinochilus
Dnd.	= Dossinodes
Dny.	= Dossinyera
Dsi.	= Dossisia
Dot.	= Dothilopsis
Dwsa.	= Downsara
Drac.	= Dracula
Drvla.	= Dracuvallia
Dres.	= Dresslerara
Dre.	= Dressleria
Dress.	= Dressleriella
Dwt.	= Drewettara
Dry.	= Drymoanthus
Duk.	= Duckittara
Dugg.	= Duggerara
Dug.	= Dungsara
Dng.	= Dungsia
Dnna.	= Dunnara
Dngra.	= Dunningara
Dnv.	= Dunstervilleara
Dup.	= Dupontara
Du.	= Durabaculum
Drt.	= Durantara
Dtya.	= Durutyara
Duv.	= Duvalara
Dvv.	= Duvivierara
Dkt.	= Dyakanthus
Dy.	= Dyakia
Eas.	= Eastonara
Ecr.	= Echinorhyncha
Eudl.	= Ecuadorella
Ecd.	= Ecuadoria
Eda.	= Edara
Edr.	= Edeara
Elsa.	= Elearethusa
Ecth.	= Elecalthusa
Elo.	= Eleorchis
Elp.	= Elepogon
Et.	= Eleutheroglossum
Eliara	= Eliara
El.	= Ellanthera
Elx.	= Eltrolexia
Etp.	= Eltroplectris
Ews.	= Elwesara
Etha.	= Elythodia
Elth.	= Elythranthera
Emb.	= Embreea
Eig.	= Emigara
Emy.	= Emilythwaitesara
Eny.	= Enanthleya
Ehl.	= Enchelia
Eyr.	= Encyarthrolia
Ect.	= Encyclarthron
E.	= Encyclia
En.	= Encyclipedium
Enl.	= Encylaelia
Ecv.	= Encylaevola
Eyy.	= Encyleyvola
Eyp.	= Encyphila
Etl.	= Encytonavola
Eyv.	= Encyvola
Ece.	= Encyvolendrum
Edl.	= Endlicherara
Ekma.	= Engkhiamara
Eng.	= Engsoonara
Epp.	= Ephippium
Ert.	= Epiarthron
Epbkl.	= Epibarkiella
Ebcl.	= Epibrascattlaelia
Ep.	= Epibrassavola
Epbns.	= Epibrassonitis
Epb.	= Epibroughtonia
Ett.	= Epicatanthe
Eth.	= Epicatarthron
Ety.	= Epicatcyclia
Ecc.	= Epicatechea
Epctn.	= Epicatonia
Epc.	= Epicattleya
Ecll.	= Epicaulaelia
Eicl.	= Epichile
Ecl.	= Epicladium
Epy.	= Epicyclia
Epdla.	= Epidella
Epi.	= Epidendrum
Epdcm.	= Epidiacrium
Emk.	= Epidominkeria

(continued)

Epd.	= Epidrobium	Eua.	= Euanthe
Erd.	= Epierstedella	Eu.	= Euarachnides
Epgl.	= Epiglottis	Ern.	= Euarthron
Epg.	= Epigoa	Eun.	= Eucatanthe
Epl.	= Epilaelia	Eutl.	= Eucatlaelia
Eplc.	= Epilaeliocattleya	Ecp.	= Eucatophila
Eplps.	= Epilaeliopsis	Elv.	= Eucattlevola
Eis.	= Epilaelopsis	Ettn.	= Eucattonia
Elva.	= Epileptovola	Euc.	= Eucentrum
Ea.	= Epileya	Euh.	= Euchiclia
Els.	= Epiliopsis	Eal.	= Euchilaelia
Epil.	= Epilopsis	Ech.	= Euchile
Emc.	= Epimicra	Eucl.	= Euclades
Epn.	= Epinidema	Euct.	= Eucycattleya
Eps.	= Epiopsis	Elh.	= Eucyclechea
Epcts.	= Epipactis	Edv.	= Eudendravola
Eph.	= Epiphaius	Eud.	= Eudevereuxara
Eil.	= Epiphila	Eln.	= Eulaelianthe
Epip.	= Epiphronitella	Euly.	= Euleya
Ephs.	= Epiphronitis	Eyt.	= Euleyarthron
Epgm.	= Epipogium	Ebd.	= Eulobidium
Ery.	= Epirhynanthe	Eucmla.	= Eulocymbidiella
Epstm.	= Epistoma	Elm.	= Eulomangis
Etv.	= Epithechavola	Euph.	= Eulophia
Etc.	= Epithechea	Eul.	= Eulophiella
Epith.	= Epithechia	Ely.	= Eulophyllum
Enn.	= Epitonanthe	Ela.	= Eulosellia
Eptn.	= Epitonia	Emg.	= Eumingoa
Epv.	= Epivola	Eup.	= Eupapilanda
Era.	= Erasanthe	Epo.	= Eupapilio
Erm.	= Eremorchis	Epr.	= Euporphyranda
Er.	= Eria	Eur.	= Eurachnis
Eric.	= Ericara	Erc.	= Eurhyncattleya
Eht.	= Ericholttumara	Eugs.	= Euryangis
Eks.	= Erikstephenstormara	Eyb.	= Euryblema
Ena.	= Erinara	Eyc.	= Eurycaulis
Ess.	= Eriopsis	Echn.	= Eurychone
Erx.	= Erioxantha	Eugcm.	= Eurygraecum
Entra.	= Ernestara	Eunps.	= Eurynopsis
Ercn.	= Erycina	Edd.	= Euthechedendrum
Erdm.	= Erydium	Eut.	= Eutonia
Eym.	= Erymesa	Euv.	= Euvola
Eyn.	= Eryumnia	Evk.	= Evakara
Esm.	= Esmenanthera	Exa.	= Exalaria
Em.	= Esmeralda	Exe.	= Exeria
Es.	= Esmeranda	Exo.	= Exochanthus
Ems.	= Esmeropsis	Fdn.	= Fadenchoda
Est.	= Esmerstylis	Ffn.	= Fadenfinanda
Esn.	= Esperonara	Ferg.	= Fergusonara
Esta.	= Estelaara	F.	= Fernandezia

(continued)

Fia.	= Fialaara		Gsrth.	= Gastrothera
Fin.	= Finetia		Gin.	= Gastruisinda
Fsh.	= Fisherara		Gtra.	= Gauntlettara
Flt.	= Fletcherara		Gav.	= Gavilea
For.	= Forbesina		Genig.	= Genus ignota
Fdca.	= Fordyceara		Gcd.	= Geoclades
Fgtra.	= Forgetara		Gdm.	= Geodorum
Frs.	= Forsterara		Gbka.	= Georgeblackara
Frn.	= Fournierara		Gcr.	= Georgecarrara
Fow.	= Fowlerara		Ggf.	= Georgefara
Flr.	= Fowlieara		Grd.	= Gerardusara
Fdk.	= Fredclarkeara		Gba.	= Gerberara
Fre.	= Fredschechterara		Ghi.	= Ghillanyara
Frda.	= Freedara		Gbz.	= Gibezara
Frz.	= Frezierara		Gi.	= Giddingsara
Fri.	= Friedaara		Gigara	= Gigara
Frb.	= Froebelara		Gil.	= Gilmourara
Frt.	= Fruticicola		Glya.	= Gladysyeeara
Fcr.	= Fuchsara		Glz.	= Glanzara
Fjo.	= Fujioara		Glc.	= Glicensteinara
Fjw.	= Fujiwarara		Gls.	= Glossadenia
Flc.	= Fulaichangara		Gloss.	= Glossodia
Gab.	= Gabertia		Gfa.	= Goffara
Gbs.	= Galabstia		Ghta.	= Gohartia
Gal.	= Galeandra		Gld.	= Goldnerara
Gslla.	= Galeansellia		Glm.	= Golumnia
Gds.	= Galeodes		Gmda.	= Gomada
Gle.	= Galeomenetalum		Gmd.	= Gomadachtia
Gln.	= Galeonisia		Gbt.	= Gombrassiltonia
Gptm.	= Galeopetalum		Ger.	= Gomcentridium
Glta.	= Galeottia		Gdt.	= Gomcidettia
Gzl.	= Galiczella		Gcn.	= Gomcidumnia
Ga.	= Garayara		Gmk.	= Gomenkoa
Gpp.	= Garlippara		Gom.	= Gomesa
Gsta.	= Gastisia		Gcg.	= Gomesochiloglossum
Gscpa.	= Gastisocalpa		Gsc.	= Gomesochilum
Gthe.	= Gastocalanthe		Gms.	= Gomestele
Gaph.	= Gastophaius		Gmtta.	= Gomettia
Gach.	= Gastorchis		Gmz.	= Gomezina
Gsd.	= Gastranda		Gmg.	= Gomguezia
Gtts.	= Gastritis		Gtd.	= Gomiltidium
Gas.	= Gastrocalanthe		Gzn.	= Gomiltlauzina
Gchgl.	= Gastrochiloglottis		Gml.	= Gomiltostele
Gchls.	= Gastrochilus		Gmt.	= Gomocentrum
Gnp.	= Gastronopsis		Gmch.	= Gomochilus
Gpi.	= Gastrophaianthe		Gmgm.	= Gomoglossum
Gp.	= Gastrophaius		Goa.	= Gomonciada
Gs.	= Gastrorchis		Gch.	= Gomoncidochilum
Gsarco.	= Gastrosarcochilus		Gmn.	= Gomonia
Gstm.	= Gastrostoma		Gnd.	= Gomoniopcidium

(continued)

(continued)

Hnp.	= Haraenopsis		Hf.	= Hofmeisterella
Hrdg.	= Hardingia		Hct.	= Holcanthe
Hrs.	= Harrisara		Hol.	= Holcanthera
Hart.	= Hartara		Hln.	= Holcenda
Hvy.	= Harveyara		Hctm.	= Holcocentrum
Hasgw.	= Hasegawaara		Hld.	= Holcodirea
Has.	= Hasskarlara		Hcf.	= Holcofinetia
Hat.	= Hatcherara		Holc.	= Holcoglossum
Hatt.	= Hattoriara		Hlg.	= Holcograecum
Haus.	= Hausermannara		Hop.	= Holconopsis
Haw.	= Hawaiiara		Hlp.	= Holcopsis
Hwkra.	= Hawkesara		Hr.	= Holcorides
Hknsa.	= Hawkinsara		Hls.	= Holcosia
Hyt.	= Hayata		Hla.	= Holcosianda
Hay.	= Hayataara		Hoc.	= Holcostylis
Hyw.	= Haywoodara		Hvs.	= Holcovanstylis
Hlc.	= Helcia		Hfd.	= Holfordara
Hdm.	= Helenadamsara		Hog.	= Hollingtonara
Hpla.	= Helpilia		Hlm.	= Holmara
Hmp.	= Hemipiliopsis		Holtt.	= Holttumara
Hbtra.	= Herbertara		Hon.	= Honoluluara
H.	= Herminium		Hook.	= Hookerara
Hro.	= Hermoara		Hos.	= Hoosierara
Hrm.	= Hermorchis		Horm.	= Hormidium
Hrml.	= Heromeulenia		Hrn.	= Hornara
Hrpo.	= Heropaludorchis		Hhp.	= Houhopea
Hroc.	= Herorchis		Hlt.	= Houlletia
Hrp.	= Herorchiserapias		Hul.	= Houllinia
Hrr.	= Herreraara		Hlra.	= Houllora
Hers.	= Herschelia		How.	= Howardara
Hrds.	= Herscheliodisa		Hwra.	= Howeara
Hrt.	= Hertensteinara		Hru.	= Hrubyara
Hts.	= Heterotaxis		Hsu.	= Hsuara
Htz.	= Heterozeuxine		Hng.	= Huangara
Hex.	= Hexadesmia		Hylra.	= Hueylihara
Hxsa.	= Hexisea		Hgfda.	= Hugofreedara
Hyn.	= Heynholdara		Hmb.	= Humboldtara
Hiat.	= Hiattara		Humm.	= Hummelara
Hgsh.	= Higashiara		Hu.	= Huntara
Hig.	= Higginsara		Htg.	= Huntingtonara
Hdra.	= Hildaara		Hnths.	= Huntleanthes
Him.	= Himantoglossum		Hya.	= Huntleya
Hmra.	= Himoriara		Huz.	= Huntzellanthes
Hry.	= Hirayamaara		Hzl.	= Hunzella
His.	= Hispaniella		Hur.	= Hurstara
Hhn.	= Hoehneara		Hybo.	= Hybriorchis
Hfc.	= Hoffmanncattleya		Hyr.	= Hydranthus
Hfy.	= Hoffmanncyclia		Hyea.	= Hyeara
Hof.	= Hoffmannseggella		Hcm.	= Hyedecromara
Hfm.	= Hofmannara		Hyd.	= Hygranda

(continued)

Hcd.	= Hygrocenda		Jkl.	= Johnkellyara
Hy.	= Hygrochilus		Jol.	= Johnlagerara
Hgd.	= Hygrodirea		Joh.	= Johnsonara
Hmn.	= Hymeneria		Jya.	= Johnyeeara
Hmc.	= Hymenochilus		Jly.	= Jolyara
Icvl.	= Iacovielloara		Jmk.	= Jomkhwanara
Ian.	= Ianara		Jks.	= Jomkhwansaeleeara
Ink.	= Ianclarkara		Jsc.	= Jonesiopchis
Ibi.	= Ibidium		Joe.	= Jonesiopsis
Icj.	= Ichijoara		Jnc.	= Jonorchis
Ich.	= Ichthyostomum		Jsp.	= Josephara
Ida	= Ida		Jos.	= Jostia
Ilo.	= Ilonara		Jmth.	= Jumanthes
Ing.	= Ingramara		Jum.	= Jumellea
Ino.	= Inobulbon		Jng.	= Jungara
In.	= Inti		Kgw.	= Kagawaara
I.	= Ioncidium		Kal.	= Kalakauara
Intta.	= Ionettia		K.	= Kamemotoara
Ims.	= Ionmesa		Kun.	= Kanetsunaara
Imt.	= Ionmesettia		Kza.	= Kanzerara
Ict.	= Ionocentrum		Kar.	= Karorchis
Incdm.	= Ionocidium		Kat.	= Katherinea
Iod.	= Ionomesidium		Kts.	= Kautskyara
Inps.	= Ionopsis		Kwmta.	= Kawamotoara
Ipt.	= Ionparettichilum		Kaw.	= Kawanishiara
Inm.	= Ionumnia		Kefth.	= Keferanthes
Ip.	= Ipsea		Kfl.	= Keferella
Ips.	= Ipseglottis		Kfy.	= Keferhyncha
Ire.	= Irenea		Kfz.	= Kefericzella
Irv.	= Irvingara		Krl.	= Keferollea
Isa.	= Isabelia		Kefst.	= Kefersteinia
Isd.	= Isadendrum		Kfr.	= Keforia
Is.	= Isanitella		Kft.	= Keftorella
Isr.	= Isaoara		Ke.	= Kegeliella
Iwan.	= Iwanagaara		Kgp.	= Kegeopea
Iana.	= Iwanagara		Kei.	= Keishunara
Ixy.	= Ixyophora		Kht.	= Keishunhattoriara
Izma.	= Izumiara		Knb.	= Kennethbealeara
Jkf.	= Jackfowlieara		Ker.	= Kerchoveara
Jqn.	= Jacquinara		Key.	= Keyesara
J.	= Jacquinparis		Khm.	= Khiamara
Jmr.	= Jamesonara		Ktn.	= Kiattanara
Jan.	= Janssensara		Kim.	= Kimballara
Jen.	= Jeaneara		Ki.	= Kingidium
Jel.	= Jellesmaara		King.	= Kingiella
Jwa.	= Jewellara		Kgs.	= Kingistylis
Jmzra.	= Jimenezara		Kin.	= Kionophyton
Jsr.	= Jisooara		Kpa.	= Kippenara
Jnna.	= Joannara		Kir.	= Kirchara
Jon.	= Johnara		Kit.	= Kitigorchis

(continued)

Klma.	= Klehmara
Klg.	= Klugara
Knp.	= Knappara
Knw.	= Knowlesara
Knd.	= Knudsenara
Knud.	= Knudsonara
Kdm.	= Kodamaara
Kmv.	= Komarovara
Kom.	= Komkrisara
Kro.	= Kraenlinorchis
Kr.	= Kraenzlinara
Kz.	= Kraenzlinella
Krsa.	= Kraussara
Kgra.	= Kriegerara
Kn.	= Kuhnara
Kno.	= Kunoara
Knt.	= Kunthara
Lcl.	= Lachelinara
Lah.	= Laechilis
Lga.	= Laegoa
Lctt.	= Laelcattleya
L.	= Laelia
Lbc.	= Laelia-Brasso-Cattleya
Lnt.	= Laelianthe
Lvl.	= Laeliavola
Lch.	= Laelichilis
Ldt.	= Laelidendranthe
Lcn.	= Laeliocatanthe
Lcr.	= Laeliocatarthron
Lctna.	= Laeliocatonia
Lcka.	= Laeliocattkeria
Lc.	= Laeliocattleya
Ldrn.	= Laeliodendron
La.	= Laeliodendrum
Lkra.	= Laeliokeria
Lpya.	= Laeliopleya
Lps.	= Laeliopsis
Lv.	= Laeliovola
Lrn.	= Laelirhynchos
Lna.	= Laelonia
Lpn.	= Laenopsonia
Laeo.	= Laeopis
Lae.	= Laeopsis
Lca.	= Laerianchea
Ltt.	= Laetonanthe
Lgra.	= Lagerara
Lpca.	= Laipenchihara
Lair.	= Lairesseara
Lmb.	= Lambara
Lam.	= Lambeauara
Lbka.	= Lancebirkara
Lauara	= Lauara
Lav.	= Lavrihara
Lws.	= Law-Schofieldara
Lwr.	= Lawara
Law.	= Lawlessara
Lwn.	= Lawrenceara
Lay.	= Laycockara
Lnya.	= Leaneyara
Leb.	= Lebaudyara
Led.	= Ledienara
Leeara	= Leeara
Leh.	= Leechara
Lee.	= Leemannara
Lei.	= Leioanthum
Lemra.	= Lemaireara
Lem.	= Lemboglossum
Llm.	= Leochilumnia
Lchs.	= Leochilus
Lcdm.	= Leocidium
Lcmsa.	= Leocidmesa
Lcdpa.	= Leocidpasia
Lod.	= Leocidumnia
Lgn.	= Leogolumnia
Lko.	= Leokoa
Lsz.	= Leomesezia
Len.	= Leonara
Lep.	= Lepanopsis
Lths.	= Lepanthes
Lpths.	= Lepanthopsis
Lptdm.	= Leptodendrum
Lgt.	= Leptoguarianthe
Lptka.	= Leptokeria
Lptl.	= Leptolaelia
Lpt.	= Leptotes
Lptv.	= Leptovola
Lesl.	= Leslieara
Lhr.	= Lesliehertensteinara
Lsu.	= Lesueurara
Lth.	= Letochilum
Lce.	= Leucohyle
Le.	= Leucorchis
Lev.	= Levyara
Lwsra.	= Lewisara
Liaps.	= Liaopsis
Licht.	= Lichtara
Lich.	= Lichterveldia
Lieb.	= Liebmanara
Lim.	= Limara
Lima.	= Limatopreptanthe

(continued)

Lnr.	= Linara	Lut.	= Luisanthera
Lblm.	= Lindblomia	Lda.	= Luisedda
Lin.	= Lindleyara	Luiser.	= Luiserides
Lya.	= Lindleyella	Lsa.	= Luisia
Lgl.	= Linguella	Lups.	= Luisiopsis
Linn.	= Linneara	Lst.	= Luistylis
Lpna.	= Lioponia	Lvta.	= Luivanetia
Lis.	= Listera	Luth.	= Lutherara
Lob.	= Lobbara	Lbs.	= Lycabstia
Lkcdm.	= Lockcidium	Lfl.	= Lycafrenuloa
Lkda.	= Lockcidmesa	Lmc.	= Lycamerlycaste
Lhta.	= Lockhartia	Lns.	= Lycanisia
Lck.	= Lockia	Lyc.	= Lycaste
Lkctta.	= Lockochilettia	Ly.	= Lycastenaria
Lkchs.	= Lockochilus	Lct.	= Lycastiella
Lkg.	= Lockoglossum	Lzl.	= Lycazella
Lkgch.	= Lockogochilus	Lcd.	= Lycida
Lckp.	= Lockopilia	Lyb.	= Lycobyana
Lkstx.	= Lockostalix	Lymra.	= Lymanara
Lkm.	= Lockumnia	Lyon.	= Lyonara
Lts.	= Loefgrenianthus	Lci.	= Lyonarci
Ltl.	= Lomantrisuloara	Lrd.	= Lyridium
Lon.	= Londesboroughara	Lys.	= Lysudamuloa
Lgh.	= Longhueiara	Mcb.	= Macbrideara
Lop.	= Lophiaris	Mcq.	= Maccorquodaleara
Loph.	= Lophoglottis	Mcyra.	= Maccoyara
Lora.	= Lorenara	Mcc.	= Maccraithara
Lrt.	= Loritis	Mrt.	= Maccraithea
Lor.	= Loroglorchis	Mly.	= Maccullyara
Lou.	= Louiscappeara	Mcdl.	= Macdanielara
Lov.	= Lovelessara	Maka.	= Macekara
Low.	= Lowara	Mclna.	= Maclellanara
Lwc.	= Lowiorchis	Mclmra.	= Maclemoreara
Lwnra.	= Lowsonara	Mmk.	= Macmeekinara
Lwt.	= Lowsutongara	Mac.	= Macodes
Lscta.	= Luascotia	Mcd.	= Macodisia
Luc.	= Lucasara	Mdy.	= Macodyera
Lus.	= Ludisia	Mcdn.	= Macradenia
Ldw.	= Ludlowara	Mcdsa.	= Macradesa
Lud.	= Ludochilus	M.	= Macrangraecum
Lnpt.	= Luianopsanthe	Mstn.	= Macrastonia
Lctm.	= Luicentrum	Mpt.	= Macropodanthus
Luic.	= Luichilus	Mur.	= Macrura
Lnd.	= Luilionanda	Mdo.	= Maderoara
Lnta.	= Luinetia	Mta.	= Maechtleara
Lpd.	= Luinopsanda	Mae.	= Maelenia
Lnps.	= Luinopsis	Mai.	= Mailamaiara
Lid.	= Luiphalandopsis	Mcba.	= Malcolmcampbellara
Lu.	= Luisaerides	Mtn.	= Mantinara
Lsnd.	= Luisanda	Mpn.	= Mapinguari

(continued)

Map.	= Mapingucaste	Mst.	= Microstylis
Mph.	= Marcchristopherstormara	Mty.	= Microthelys
Mrm.	= Marimerara	Msr.	= Milassentrum
Mrn.	= Maronara	Mlc.	= Milcentrum
Mrr.	= Marriottara	Mis.	= Milcidossum
Msh.	= Marshara	Mkd.	= Milenkocidium
Mtr.	= Martiusara	Mla.	= Millerara
Mrv.	= Marvingerberara	Msp.	= Millspaughara
Mry.	= Marycrawleystormara	Mmc.	= Milmilcidium
Masd.	= Masdevallia	Mmo.	= Milmiloda
Msna.	= Masonara	Mmg.	= Milmiloglossum
Mrp.	= Masrepia	Mmp.	= Milmilpasia
Msg.	= Massangeara	Mmr.	= Milmilrassia
Mg.	= Mastigion	Mmt.	= Milmiltonia
Mtw.	= Mathewsara	Mlr.	= Miloara
Msda.	= Matsudaara	Moz.	= Milonzina
Mau.	= Maumeneara	Mpsa.	= Milpasia
Mnda.	= Maunderara	Mpla.	= Milpilia
Mrc.	= Mauriceara	Mtad.	= Miltada
Mxd.	= Maxidium	Mtadm.	= Miltadium
Mxcst.	= Maxillacaste	Mtta.	= Miltarettia
Max.	= Maxillaria	Mtssa.	= Miltassia
Mxl.	= Maxillariella	Mtst.	= Miltistonia
Mxy.	= Maxillyca	Mld.	= Miltochilidium
Mxlb.	= Maxilobium	Mtc.	= Miltochilum
Mxp.	= Maxthompsonara	Mtd.	= Miltodontrum
May.	= Mayara	Mgc.	= Miltogomechilum
Mymra.	= Maymoirara	Mnt.	= Miltoncentrum
Mea.	= Meadara	Mcid.	= Miltoncidium
Msu.	= Measuresara	Mos.	= Miltoncidostele
Med.	= Mediocalcar	Mz.	= Miltonguezia
Mchr.	= Meechaiara	Milt.	= Miltonia
Mrclm.	= Meiracyllium	Mil.	= Miltonicidium
Mei.	= Meirmosesara	Mtdm.	= Miltonidium
Mel.	= Meloara	Mtda.	= Miltonioda
Mdl.	= Mendelara	Mps.	= Miltoniopsis
Mdcla.	= Mendoncella	Mp.	= Miltonpasia
Mzr.	= Menziesara	Mpa.	= Miltonpilia
Mdla.	= Mesadenella	Msl.	= Miltostelada
Meh.	= Meshaara	Mrd.	= Mirandorchis
Mes.	= Mesoglossum	Mzu.	= Mizunoara
Meo.	= Mesospinidium	Miz.	= Mizutara
Met.	= Metdepenningenara	Moe.	= Moensara
Mvt.	= Michelvacherotara	Mhw.	= Moihwaara
Mchza.	= Micholitzara	Moir.	= Moirara
Mit.	= Microcattleya	Mkra.	= Mokara
Mcr.	= Microchilus	Mnkr.	= Monkara
Mpd.	= Microepidendrum	Mkhsa.	= Monkhouseara
Mcl.	= Microlaelia	Monn.	= Monnierara
Micr.	= Micropera	Mnra.	= Moonara

(continued)

Mora.	= Mooreara		Neoj.	= Neojoannara
Moi.	= Morieara		Ng.	= Neokagawara
Mrml.	= Mormariella		Neom.	= Neomoirara
Mor.	= Mormleria		Nk.	= Neomokara
Morm.	= Mormodes		Npp.	= Neopabstopetalum
Mo.	= Mormodia		Nb.	= Neorobinara
Mlca.	= Mormolyca		Ndd.	= Neosedanda
Mml.	= Mormosellia		Nsd.	= Neosedirea
Mrsa.	= Morrisonara		Neost.	= Neostylis
Mscra.	= Moscosoara		Nsls.	= Neostylopsis
Mtf.	= Mountfordara		Ntn.	= Neotainiopsis
Mrth.	= Mycaranthes		Ntc.	= Neotinacamptis
Mym.	= Mylamara		Ntz.	= Neotinarhiza
My.	= Myoxanthus		Nt.	= Neotinea
Mxt.	= Myoxastrepia		Neot.	= Neottia
Mcn.	= Myrmecanthe		Nyf.	= Neoyusofara
Mcv.	= Myrmecatavola		Nzg.	= Neozygisia
Mycl.	= Myrmecatlaelia		Ngara	= Ngara
Myv.	= Myrmecavola		Nls.	= Nicholasara
Myh.	= Myrmechea		Ncl.	= Nicholsonara
Myc.	= Myrmecocattleya		Nkz.	= Nickcannazzaroara
Mco.	= Myrmecochile		Ndc.	= Nideclia
Myl.	= Myrmecolaelia		Nid.	= Nidema
Mcp.	= Myrmecophila		Npn.	= Nipponorchis
Myp.	= Myrmeopsis		Nlra.	= Nobleara
Mys.	= Myrmesophleya		Nct.	= Nochocentrum
Myt.	= Myrmetonia		Nog.	= Nogomesa
Mycdm.	= Mystacidium		Ndp.	= Nohacidiopsis
Mro.	= Mystorchis		Nhc.	= Nohacidium
Ngl.	= Nageliella		Ngk.	= Nohagomenkoa
Nkgwa.	= Nakagawaara		Nhl.	= Nohalumnia
Nak.	= Nakamotoara		Nmo.	= Nohamiltocidium
Nkm.	= Nakamuraara		Nmt.	= Nohamiltonia
Nash.	= Nashara		Nhp.	= Nohamiltoniopsis
Naug.	= Naugleara		Nsm.	= Nohastelomesa
Neb.	= Nebrownara		Nwk.	= Nohawenkoa
Npg.	= Neippergia		Ntr.	= Nohawilentrum
Nrst.	= Neoaeristylis		Nhw.	= Nohawilliamsia
Nbth.	= Neobathiea		Nzc.	= Nohazelencidium
Nbps.	= Neobatopus		Non.	= Nonaara
Nbsa.	= Neobolusia		Nhmta.	= Nornahamamotoara
Nca.	= Neochristieara		Nrna.	= Northenara
Neo.	= Neodebruyneara		Nwda.	= Norwoodara
Nex.	= Neodevereuxara		No.	= Nothodoritis
Ndn.	= Neofadanda		Ntd.	= Notolidium
Nfd.	= Neofadenia		Ntlta.	= Notylettia
Neof.	= Neofinetia		Ntl.	= Notylia
Ngda.	= Neogardneria		Ntldm.	= Notylidium
Neogm.	= Neoglossum		Ntlps.	= Notylopsis
Ngrcm.	= Neograecum		Nuc.	= Nuccioara

(continued)

Oks.	= Oakesara	Onz.	= Oncidquezia
Obr.	= Obrienara	Ocd.	= Oncidumnia
Oddy.	= Oddyara	Oncigom.	= Oncigomada
Ode.	= Oderara	Otd.	= Oncitonioides
Ocp.	= Odonchlopsis	Ons.	= Oncostele
Oda.	= Odontioda	Osp.	= Oncostelopsis
O.	= Odontiodonia	Onra.	= Onoara
Otp.	= Odontiopsis	Odk.	= Onrodenkoa
Odbrs.	= Odontobrassia	Otz.	= Ontolezia
Otc.	= Odontocentrum	Ogs.	= Ontolglossum
Odt.	= Odontochilus	Olt.	= Onzelcentrum
Odcdm.	= Odontocidium	Ozt.	= Onzelettia
Odm.	= Odontoglossum	Ozl.	= Onzeloda
Otk.	= Odontokoa	Ozn.	= Onzelumnia
Odtna.	= Odontonia	Opt.	= Ophramptis
Odpla.	= Odontopilia	Oph.	= Ophrys
Odrta.	= Odontorettia	Opisan.	= Opisanda
Ots.	= Odontostele	Opo.	= Opoixara
Otl.	= Odontozelencidium	Ops.	= Opsilaelia
Od.	= Odopetalum	Opsis.	= Opsisanda
Ody.	= Odyncidium	Op.	= Opsisanthe
Oecl.	= Oeceoclades	Opsct.	= Opsiscattleya
Oeo.	= Oeonia	Opst.	= Opsistylis
Oenla.	= Oeoniella	Orc.	= Orchidactyla
Oec.	= Oeorchis	Oy.	= Orchidactylorhiza
Oe.	= Oerstedella	Ogy.	= Orchigymnadenia
Ork.	= Oerstedkeria	Org.	= Orchimantoglossum
Osl.	= Oerstelaelia	Ohn.	= Orchinea
Oer.	= Oertonia	Orcys.	= Orchiophrys
Ost.	= Oestlundia	Opa.	= Orchiplatanthera
Okr.	= Okaara	Orchis	= Orchis
Olg.	= Olgasis	Orsps.	= Orchiserapias
Ogt.	= Oligochaetochilus	Ore.	= Oreorchis
Ora.	= Oncandra	Orcp.	= Ornithocephalus
Ont.	= Oncidaretia	Orncm.	= Ornithocidium
Ott.	= Oncidarettia	Orpha.	= Ornithophora
Osa.	= Oncidasia	Orp.	= Orpetara
Oncna.	= Oncidenia	Osmt.	= Osmentara
Oncsa.	= Oncidesa	Ogm.	= Osmoglossum
Onctta.	= Oncidettia	Otr.	= Otaara
Oncg.	= Oncidguezia	Obn.	= Otobrastonia
Onclla.	= Oncidiella	Otcd.	= Otocidium
Oncda.	= Oncidioda	Otcx.	= Otocolax
Onc.	= Oncidium	Oth.	= Otoglochilum
Odd.	= Oncidodontopsis	Oto.	= Otoglossum
Oig.	= Oncidoglossum	Otht.	= Otohartia
Onp.	= Oncidophora	Otnsa.	= Otonisia
Odp.	= Oncidopsiella	Ot.	= Otopabstia
Oip.	= Oncidopsis	Ohd.	= Otorhynchocidium
Oncpa.	= Oncidpilia	Otspm.	= Otosepalum

(continued)

Oot.	= Otostele		Phcg.	= Paraholcoglossum
Otst.	= Otostylis		Pph.	= Paralophia
Ota.	= Ottoara		Py.	= Paramayara
Owsr.	= Owensara		Pas.	= Parandachnis
Oxy.	= Oxyglossellum		Pdt.	= Parandanthe
Oys.	= Oxysepala		Pn.	= Paranthe
Pbn.	= Pabanisia		Prt.	= Paranthera
Pabs.	= Pabstara		Ptt.	= Paraottis
Pab.	= Pabstia		Ppl.	= Parapapilio
Pss.	= Pabstosepalum		Pphc.	= Paraphachilus
Pga.	= Pageara		Pdn.	= Paraphadenia
Pal.	= Palermoara		Pps.	= Paraphalaenopsis
Plmra.	= Palmerara		Pplt.	= Paraphalanthe
Pldm.	= Paludomeulenia		Prec.	= Paraphalraecum
Pdo.	= Paludorchis		Pts.	= Parapteroceras
Pldp.	= Paludorchiserapias		Ppt.	= Paraptosiella
Pua.	= Palumbina		Pcs.	= Pararachnis
Pana.	= Panarica		Prn.	= Pararenanthera
Pzka.	= Panczakara		Prd.	= Pararides
Pntp.	= Pantapaara		Psarco.	= Parasarcochilus
Pna.	= Paphinia		Pst.	= Parastylis
Phnp.	= Paphinopea		Pv.	= Paravanda
Paph.	= Paphiopedilum		Prv.	= Paravandaenopsis
Pa.	= Papilachnis		Paravand.	= Paravandanthe
Peps.	= Papilaenopsis		Pav.	= Paravandanthera
Psts.	= Papilaenostylis		Pvp.	= Paravandopsis
Pap.	= Papilandachnis		Pvd.	= Paravandrum
Pl.	= Papilanthera		Pia.	= Parisia
Pc.	= Papiliocentrum		Pka.	= Parkerara
Pd.	= Papiliodes		Parn.	= Parnataara
Pda.	= Papilionanda		Pta.	= Pattonia
Ple.	= Papilionanthe		Pat.	= Pattoniheadia
Pt.	= Papilionetia		Pdr.	= Paulandmarystormara
Pp.	= Papiliopsis		Plra.	= Paulara
Pio.	= Papiliosarcanthopsis		Plsra.	= Paulsenara
Plv.	= Papiliovanvanda		Put.	= Paulstormara
Papi.	= Papilisia		Pvn.	= Pavonara
Pbm.	= Papillilabium		Pay.	= Paynterara
Pic.	= Papillochilus		Pasr.	= Peaseara
Ptn.	= Papitonia		Pe.	= Pectabenaria
Pops.	= Papopsis		Pee.	= Peetersara
Plp.	= Papulipetalum		Peh.	= Pehara
Part.	= Paracentrum		Plctm.	= Pelacentrum
Prcls.	= Parachilus		Pelcs.	= Pelachilus
Par.	= Parachnis		Pelst.	= Pelastylis
Pds.	= Paradisanisia		Pelt.	= Pelatanda
Pdsnth.	= Paradisanthus		Pthia.	= Pelatantheria
Pis.	= Paradistia		Petp.	= Pelathanopsis
Prf.	= Parafinetia		Pltrs.	= Pelatoritis
Pgnt.	= Paragnathis		Pel.	= Pelexia

(continued)

Pdlt.	= Pendletonara	Pmc.	= Phymatochilum
Pnl.	= Pennellara	Psg.	= Physogyne
Pent.	= Pentisea	Pys.	= Physothallis
Ppa.	= Pepeara	Pier.	= Pierardia
Perths.	= Peristeranthus	Pilm.	= Pilumna
Prschs.	= Peristerchilus	Pina.	= Pinalia
Per.	= Peristeria	Pi.	= Pittara
Prra.	= Perreiraara	P.	= Platanthera
Perr.	= Perrierara	Pns.	= Platyclinis
Peb.	= Pescabstia	Plchs.	= Plectochilus
Pcy.	= Pescantleya	Plrhz.	= Plectorrhiza
Pyha.	= Pescarhyncha	Plgcm.	= Plectrelgraecum
Psbol.	= Pescatobollea	Plmths.	= Plectrelminthus
Pestea.	= Pescatorea	Pln.	= Pleione
Pes.	= Pescatoria	Plnl.	= Pleionilla
Pcp.	= Pescatoscaphe	Pll.	= Plelis
Psw.	= Pescawarrea	Prm.	= Pleurobotryum
Pesc.	= Pescenia	Pths.	= Pleurothallis
Psnth.	= Pescoranthes	Plm.	= Plumatichilos
Pmr.	= Pesmanara	Pmt.	= Plumatistylis
Pth.	= Peterhuntara	Ppg.	= Poeppigara
Ptr.	= Petrorchis	Poll.	= Pollardia
Pett.	= Pettitara	Plet.	= Pollettara
Ph.	= Phabletia	Pcn.	= Polycycnis
Phcal.	= Phaiocalanthe	Plr.	= Polygora
Phcym.	= Phaiocymbidium	Ppx.	= Polyphylax
Phai.	= Phaiolimatopreptanthe	Prad.	= Polyradicion
Ppp.	= Phaiopreptanthe	Pol.	= Polystachya
Phaius	= Phaius	Pmctm.	= Pomacentrum
Plgs.	= Phalaeglossum	Pom.	= Pomanda
Pht.	= Phalaenetia	Pmtsa.	= Pomatisia
Phd.	= Phalaenidium	Pmcpa.	= Pomatocalpa
Php.	= Phalaenopapilio	Pmtls.	= Pomatochilus
Phal.	= Phalaenopsis	Pnr.	= Ponerorchis
Phns.	= Phalaensonia	Ptg.	= Ponerostigma
Phda.	= Phalaerianda	Pnt.	= Ponterara
Phdps.	= Phalandopsis	Ptva.	= Ponthieva
Phnta.	= Phalanetia	Polra.	= Pooleara
Pte.	= Phalanthe	Por.	= Porphyrachnis
Pld.	= Phaleralda	Pra.	= Porphyranda
Phlla.	= Phaliella	Prs.	= Porphyrandachnis
Pvv.	= Phalvanvanda	Prtr.	= Porphyranthera
Phc.	= Pharochilum	Po.	= Porphyrodesme
Phel.	= Pheladenia	Pgt.	= Porphyroglottis
Phl.	= Philippiara	Porglot.	= Porphyroglottis invalid
Phill.	= Phillipsara	Poh.	= Porphyrorhynchos
Phm.	= Phragmipaphiopedilum	Porp.	= Porphyrostachys
Phrphm.	= Phragmipaphium	Pcu.	= Porracula
Phrag.	= Phragmipedium	Prgm.	= Porroglossum
Phragmo.	= Phragmopaphium	Pvla.	= Porrovallia

(continued)

Prta.	= Porterara
Poi.	= Portillia
Port.	= Portoara
Pad.	= Posadaara
Pot.	= Potinara
Pou.	= Pourbaixara
Pow.	= Powellara
Pdhn.	= Pradhanara
Prp.	= Prapinara
Prpc.	= Preptacalanthe
Pr.	= Preptanthe
Pri.	= Priceara
Pbr.	= Probaranthe
Prte.	= Procaste
Pcv.	= Procatavola
Poha.	= Prochaea
Poc.	= Proctoria
Pcc.	= Procycleya
Pdd.	= Prodendranthe
Pgy.	= Proguarleya
Plh.	= Proleyophila
Pre.	= Proleytonia
Pmds.	= Promadisanthus
Pmar.	= Promarrea
Pcd.	= Promcidium
Pmla.	= Promellia
Pmb.	= Promenabstia
Prom.	= Promenaea
Prths.	= Promenanthes
Pmp.	= Promenopsis
Pmz.	= Promenzella
Pgl.	= Promoglossum
Prsm.	= Promosepalum
Pbt.	= Propabstopetalum
Pop.	= Propescapetalum
Pptm.	= Propetalum
Pnp.	= Prosanthopsis
Prh.	= Prosarthron
Psv.	= Prosavola
Psr.	= Proscatarthron
Prch.	= Proschile
Pg.	= Prosgoa
Plc.	= Proslaeliocattleya
Psl.	= Proslia
Ppv.	= Prosophrovola
Pry.	= Prosrhyncholeya
Psh.	= Prosthechea
Pros.	= Prostonia
Prc.	= Prosyclia
Ps.	= Pseudadenia
Psd.	= Pseudanthera
Pde.	= Pseudencyclia
Pdm.	= Pseudinium
Pcg.	= Pseudocoeloglossum
Pdh.	= Pseudohemipilia
Pdla.	= Pseudolaelia
Pse.	= Pseudorchis
Psz.	= Pseudorhiza
Psil.	= Psilanthemum
Pbc.	= Psybrassocattleya
Pbg.	= Psyburgkia
Pyrt.	= Psycarthron
Psct.	= Psycattleytonia
Pyv.	= Psycavola
Phh.	= Psychanthe
Phs.	= Psychassia
Pye.	= Psychelia
Psyh.	= Psychia
Psy.	= Psychilis
Pyn.	= Psychlumnia
Pyc.	= Psychocentrum
Pyd.	= Psychocidium
Pyl.	= Psychoglossum
Pms.	= Psychomesa
Pyo.	= Psychophila
Psp.	= Psychopilia
Psyc.	= Psychopsiella
Pyp.	= Psychopsis
Ppy.	= Psychopsychopsis
Pyct.	= Psylaeliocattleya
Ply.	= Psyleyopsis
Pym.	= Psymiltonia
Pso.	= Psyonitis
Ppc.	= Psysophrocattleya
Pyh.	= Psythechea
Pyt.	= Psytonia
Pos.	= Pteroceras
Ptc.	= Pterocottia
Pgsa.	= Pteroglossa
Ptx.	= Pterolexia
Ptp.	= Pteroplodium
Ptst.	= Pterostylis
Pgdm.	= Pterygodium
Pur.	= Purverara
Pya.	= Pynaertara
Pyr.	= Pyrorchis
Qrk.	= Quirkara
Qvl.	= Quisavola
Qch.	= Quischilis
Qui.	= Quisqueya

(continued)

Q.	= Quisumbingara
Rdnc.	= Radinocion
Raf.	= Rafinesqueara
Rgn.	= Raganara
Rmsya.	= Ramasamyara
R.	= Randactyle
Rgs.	= Rangaeris
Ran.	= Ranorchis
Rap.	= Rappartara
Rau.	= Rauhara
Reb.	= Rebeccaara
Recc.	= Recchara
Recchi.	= Recchiara
Rchg.	= Rechingerara
Reg.	= Regnierara
Reh.	= Rehderara
Rfda.	= Rehfieldara
Rec.	= Reicheara
Rei.	= Reichenbachara
Rk.	= Reinikkaara
Rcl.	= Renachilus
Rnds.	= Renades
Rfnda.	= Renafinanda
Rngl.	= Renaglottis
Rnctm.	= Renancentrum
Re.	= Renanda
Rnet.	= Renanetia
Rnps.	= Renanopsis
Rpd.	= Renanparadopsis
Rnst.	= Renanstylis
Rntda.	= Renantanda
Ren.	= Renanthera
Rta.	= Renantheranda
Rena.	= Renanthoceras
Rngm.	= Renanthoglossum
Rnthps.	= Renanthopsis
Rd.	= Renaradorum
Rdm.	= Renarodorum
Rnt.	= Renata
Rpr.	= Renoprora
Rps.	= Renopsis
Rpc.	= Renorphorchis
Rvv.	= Renvanvanda
Ret.	= Restesia
Rstp.	= Restrepia
Rhenanth.	= Rhenanthopsis
Rhin.	= Rhinerrhiza
Rrh.	= Rhinerrhizochilus
Rrp.	= Rhinerrhizopsis
Rhincs.	= Rhinochilus
Rhc.	= Rhinocidium
Rdg.	= Rhipidangis
Rhip.	= Rhipidoglossum
Rh.	= Rhizanthera
Rhdm.	= Rhodehamelara
Rhm.	= Rhomboda
Ryp.	= Rhycopelia
Ryy.	= Rhyleyaopsis
Rln.	= Rhynaelionitis
Rry.	= Rhynarthrolyea
Rrt.	= Rhynarthron
Rbg.	= Rhynburgkia
Rya.	= Rhyncada
Rds.	= Rhyncadamesa
Ryn.	= Rhyncanthe
Rcc.	= Rhyncatclia
Rnd.	= Rhyncatdendrum
Ryc.	= Rhyncatlaelia
Rth.	= Rhyncattleanthe
Rym.	= Rhynchamsia
Ryh.	= Rhynchanthe
Rvm.	= Rhynchavolarum
Rby.	= Rhynchobrassoleya
Rhctm.	= Rhynchocentrum
Rch.	= Rhynchochile
Ryl.	= Rhynchochilopsis
Rych.	= Rhynchochilus
Rdd.	= Rhynchodendrum
Rcn.	= Rhynchodenia
Rdt.	= Rhynchodenitis
Ryd.	= Rhynchodirea
Rfd.	= Rhynchofadanda
Rgl.	= Rhynchoguarlia
Rlb.	= Rhyncholabium
Rl.	= Rhyncholaelia
Rlc.	= Rhyncholaeliocattleya
Rmd.	= Rhynchomesidium
Rmy.	= Rhynchomyrmeleya
Rnc.	= Rhynchonia
Rhnps.	= Rhynchonopsis
Ry.	= Rhynchopapilisia
Rnp.	= Rhynchopera
Rycp.	= Rhynchopsis
Rop.	= Rhynchopsyleya
Ryrt.	= Rhynchorettia
Rhrds.	= Rhynchorides
Rsc.	= Rhynchosophrocattleya
Rst.	= Rhynchostele
Rhy.	= Rhynchostylis
Rct.	= Rhynchothechea

(continued)

Rhh.	= Rhynchothechlia	Rgo.	= Rodrigoa
Rhv.	= Rhynchovanda	Rdza.	= Rodriguezia
Rv.	= Rhynchovandanthe	Rdzlla.	= Rodrigueziella
Rcv.	= Rhynchovola	Rms.	= Rodrimesastele
Rll.	= Rhynchovolaelia	Rodps.	= Rodriopsis
Rvt.	= Rhynchovolanthe	Rdtna.	= Rodritonia
Rvl.	= Rhynchovolitis	Rdc.	= Rodroncidilum
Rcm.	= Rhynchumnia	Rdr.	= Rodrostele
Rhcd.	= Rhyncidlioda	Rol.	= Rodrostelettia
Rcr.	= Rhyncleiserides	Rdo.	= Rodrostelidium
Rcd.	= Rhyncorades	Rrm.	= Rodrumnia
Rvd.	= Rhyncovanda	Rob.	= Roeblingara
Rcy.	= Rhyncyclia	Rlk.	= Roellkeara
Rdl.	= Rhyndenlia	Roz.	= Roezlara
Rdn.	= Rhyndiranda	Rna.	= Rogersonara
Rhdps.	= Rhyndoropsis	Rhla.	= Rohrlara
Rtt.	= Rhynitanthe	Rolf.	= Rolfeara
Rhn.	= Rhynitis	Rwm.	= Rolfwilhelmara
Rcg.	= Rhynochlioglossum	Rls.	= Rollissonara
Rhw.	= Rhynohawidium	Rgw.	= Rongwuara
Rns.	= Rhynopsirea	Ron.	= Ronmaunderara
Rpp.	= Rhynphalandopsis	Rnya.	= Ronnyara
Ryt.	= Rhyntheconitis	Rskra.	= Rosakirschara
Rly.	= Rhyntonleya	Rsra.	= Roseara
Rys.	= Rhyntonossum	Rsy.	= Rosscyrtodium
Ryps.	= Rhynvandopsis	Rsct.	= Rossicentrum
Rht.	= Rhytionanthos	Rsm.	= Rossimilmiltonia
Rtg.	= Rhytoniglossum	Rscp.	= Rossiochopsis
Ric.	= Richardara	Ros.	= Rossioglossum
Rcmza.	= Richardmizutaara	Rss.	= Rossiostele
Rchna.	= Richardsonara	Rsot.	= Rossitolidium
Ridl.	= Ridleyara	Rot.	= Rossitonia
Rthn.	= Rittershausenara	Rsp.	= Rossitoniopsis
Rtf.	= Robertrolfeara	Rsa.	= Rossmesa
Rbt.	= Robertsara	Rsg.	= Rossotoglossum
Rbc.	= Robicentrum	Rsk.	= Rosstuckerara
Rbf.	= Robifinetia	Roth.	= Rothara
Rbnra.	= Robinara	Rca.	= Rothschildara
Rsv.	= Robinstevensara	Rwl.	= Rothwellara
Rbq.	= Robiquetia	Rtra.	= Rotorara
Rbst.	= Robostylis	Rbl.	= Rubellia
Rcfta.	= Roccaforteara	Rbn.	= Rubenara
Rdssa.	= Rodrassia	Rud.	= Rudolfiella
Ro.	= Rodrenia	Ruz.	= Ruizara
Rdtta.	= Rodrettia	Rlla.	= Rumrillara
Rdtps.	= Rodrettiopsis	Run.	= Rundleara
Rdchs.	= Rodrichilus	Rppa.	= Ruppara
Rdcm.	= Rodricidium	Rsl.	= Russellara
Rden.	= Rodridenia	Ryg.	= Rydbergara
Rdgm.	= Rodriglossum	Sac.	= Saccanthera

(continued)

Slm.	= Saccolabium
Soi.	= Sacoila
Sgka.	= Sagarikara
Skba.	= Sakabaara
Sya.	= Sallyyeeara
Snd.	= Sanda
Sand.	= Sanderara
Sjma.	= Sanjumeara
Stll.	= Santanderella
Spla.	= Saplalaara
Sapp.	= Sappanara
Sar.	= Sarcalaenopsis
Sdd.	= Sarcandides
Sarc.	= Sarcanthopsis
Sdt.	= Sarcocadetia
Srctm.	= Sarcocentrum
Scs.	= Sarcoceras
Slh.	= Sarcochilanthe
Sarco.	= Sarcochilus
Sdi.	= Sarcodirea
Sglm.	= Sarcoglossum
Srgt.	= Sarcoglottis
Slx.	= Sarcolexia
Sran.	= Sarcomoanthus
Srnps.	= Sarconopsis
Sla.	= Sarcopapilionanda
Spd.	= Sarcopodium
Schy.	= Sarcoschistotylus
Srty.	= Sarcostachys
Srth.	= Sarcothera
Srv.	= Sarcovanda
Srdts.	= Saridestylis
Srts.	= Sartylis
Svp.	= Sarvandopanthera
Svd.	= Sarvandopsis
Stk.	= Satorkis
Satm.	= Satyrium
Sdra.	= Sauledaara
Saur.	= Saurorhynchos
Svg.	= Savageara
Say.	= Sayeria
Scg.	= Scaphingoa
Scgl.	= Scaphyglottis
Sld.	= Scelcidumnia
Scel.	= Scelochilus
Scd.	= Scelodium
Sgl.	= Sceloglossum
Sln.	= Scelonia
Slt.	= Scelorettia
Schfa.	= Schafferara

Sdw.	= Scheidweilerara
Sdl.	= Schiedeella
Slga.	= Schilligerara
Ssys.	= Schistotylus
Sch.	= Schlechterara
Sct.	= Schlechterorchis
Shl.	= Schloatara
Skb.	= Schluckebieria
Sbr.	= Schombarthron
Smbv.	= Schombavola
Sb.	= Schombletia
Sba.	= Schombobrassavola
Smbcna.	= Schombocatonia
Smbc.	= Schombocattleya
Sby.	= Schombocyclia
Smbdcm.	= Schombodiacrium
Smbep.	= Schomboepidendrum
Smbl.	= Schombolaelia
Scl.	= Schombolaeliocattleya
Sbc.	= Schombolaeliocyclia
Smlp.	= Schombolaeliopsis
Smbna.	= Schombonia
Smbts.	= Schombonitis
Sbp.	= Schombophila
St.	= Schombotonia
Schom.	= Schomburgkia
Scho.	= Schomburgkio-Cattleya
Scn.	= Schomcatanthe
Scc.	= Schomcattleya
Sll.	= Schomcaulaelia
Scty.	= Schomcaulattleya
Smh.	= Schomechea
Skr.	= Schomkeria
Smy.	= Schomleycyclia
Sco.	= Schomocattleya
Smr.	= Schomrhyncattleya
Stm.	= Schuitemania
Shk.	= Schunkeara
Swf.	= Schweinfurthara
Slpt.	= Scleropteris
Snk.	= Scolnikara
Sctt.	= Scottara
Scu.	= Scullyara
Sxa.	= Seahexa
Se.	= Sealara
Sedn.	= Sedenara
Sed.	= Sedirea
Sdr.	= Sedirisia
Sedtp.	= Sediritinopsis
Sdp.	= Sediropsis

(continued)

See.	= Seegerara		Spy.	= Sophleyclia
Sgra.	= Segerara		Srt.	= Sophranthe
Srr.	= Seibertara		Sprt.	= Sophrattlia
Snn.	= Seidenanda		Sbd.	= Sophrobardendrum
Sef.	= Seidenfadenara		Spb.	= Sophrobroanthe
Sei.	= Seidenfadenia		Sop.	= Sophrobroughtonia
Sdn.	= Seidenides		Scr.	= Sophrocatarthron
Slb.	= Selbyana		Shc.	= Sophrocatcattleya
Selen.	= Selenipanthes		Sol.	= Sophrocatlaelia
Scp.	= Selenocypripedium		Srct.	= Sophrocattlaelia
Sele.	= Selenopanthes		Sc.	= Sophrocattleya
Sngs.	= Senghasara		Scy.	= Sophrocyclia
Sry.	= Seraphrys		Srg.	= Sophrogoa
Srps.	= Serapias		Sl.	= Sophrolaelia
Ser.	= Serapicamptis		Slc.	= Sophrolaeliocattleya
Srpm.	= Serapimeulenia		Sly.	= Sophrolaeliocyclia
Sz.	= Serapirhiza		Slp.	= Sophrolaeliophila
Sgr.	= Sergioara		Spl.	= Sophroleya
Srp.	= Serpenticaulis		Sphr.	= Sophronia
Sev.	= Severinara		S.	= Sophronitis
Svl.	= Sevillaara		Srl.	= Sophrophila
Shn.	= Sheehanara		Shy.	= Sophroprosleya
Srf.	= Sherriffara		Spt.	= Sophrotes
Shgra.	= Shigeuraara		Spth.	= Sophrotheanthe
Shipm.	= Shipmanara		Sht.	= Sophrothechea
Shva.	= Shiveara		Spv.	= Sophrovola
Sidr.	= Sidranara		Spa.	= Spathoglottis
Sbt.	= Siebertara		Spp.	= Spathophaius
Sid.	= Siederella		Spc.	= Speculantha
Sgrt.	= Siegeristara		Sps.	= Spiessara
Sgdm.	= Sigmacidium		Spil.	= Spilorhiza
Sgmx.	= Sigmatostalix		Spir.	= Spiranthes
Sgmt.	= Sigmettia		Spr.	= Spruceara
Silpa.	= Silpaprasertara		Srka.	= Srisukara
Sgt.	= Simpliglottis		Staal.	= Staalara
Si.	= Singaporeara		Stac.	= Stacyara
Sjm.	= Sirjeremiahara		Std.	= Staffordara
Skp.	= Skeptrostachys		Stmra.	= Stamariaara
Slad.	= Sladeara		Stn.	= Stamnorchis
Smt.	= Smithanthe		Stb.	= Stanbreea
Sm.	= Smithara		Sfdra.	= Stanfieldara
Sbgcm.	= Sobennigraecum		Stga.	= Stangora
Sbk.	= Sobennikoffia		Stncn.	= Stanhocycnis
So.	= Sobraleya		Stan.	= Stanhopea
Sob.	= Sobralia		Shstrum.	= Stanhopeastrum
Sbl.	= Sobratilla		Sat.	= Statterara
Sbn.	= Sobrinoara		Sant.	= Staurachnanthera
Slr.	= Solanderara		Sta.	= Staurachnis
Sot.	= Sopharthron		Sr.	= Stauranda
Sha.	= Sophcychea		Sup.	= Staurandopsis

(continued)

Src.	= Staurochilus
Stgl.	= Staurochoglottis
Sp.	= Stauropsis
Sur.	= Staurovanda
Strn.	= Stearnara
Stl.	= Stelbophyllum
Ste.	= Stelis
Stlma.	= Stellamizutaara
Sbm.	= Stellilabium
Stp.	= Stellipogon
Sen.	= Stenia
Stla.	= Steniella
Szl.	= Stenizella
Snb.	= Stenobolusia
Sno.	= Stenodenella
Sngl.	= Stenoglottis
Snx.	= Stenolexia
Snp.	= Stenopetella
Spg.	= Stenopogon
Strs.	= Stenorrhynchos
Sten.	= Stenorrhynchus
Stsc.	= Stenosarcos
Sty.	= Stenotyla
Stph.	= Stephenara
Stmk.	= Stephenmonkhouseara
Spf.	= Steumpfleara
Stwt.	= Stewartara
Sck.	= Stichorkis
Stor.	= Stigmatorthos
Stlb.	= Stilbophyllum
Stil.	= Stilifolium
Sto.	= Stonia
Strm.	= Stormara
Str.	= Stricklandara
Sy.	= Stylisanthe
Syl.	= Styloglossum
Sdc.	= Sudacaste
Syc.	= Sudalycenaria
Sud.	= Sudamerlycaste
Sul.	= Sudamuloa
Smg.	= Summerangis
Sum.	= Summerhayesia
Sprra.	= Susanperreiraara
Sut.	= Sutingara
Stt.	= Suttonara
Svk.	= Svenkoeltzia
Swn.	= Swanara
Sw.	= Sweetara
Syma.	= Symmonsara
Sym.	= Symphodontioda
Symp.	= Symphodontoglossum
Sda.	= Symphodontonia
Syg.	= Symphyglossonia
Tch.	= Taeniochista
Tin.	= Tainiopsis
Tak.	= Takakiara
Tkl.	= Takulumena
Ta.	= Tanakara
Tanara	= Tanara
Tsr.	= Tansirara
Tlp.	= Tapilinopsis
Tpch.	= Taprachnis
Tar.	= Taprachthera
Tpd.	= Tapranda
Tpr.	= Tapranthera
Tph.	= Tapranthopsis
Tprn.	= Taprenopsis
Trvp.	= Taprenvandopsis
Tpn.	= Taproanthe
Tpb.	= Taprobanea
Tpg.	= Taproglottis
Tpp.	= Tapronopsis
Tppd.	= Tapropapilanda
Tplt.	= Tapropapilanthera
Tprp.	= Taproparanthera
Tpln.	= Taprophalanda
Tpt.	= Taprotrichothera
Tvd.	= Tapvandachnis
Tvr.	= Tapvandera
Tat.	= Tateara
Tur.	= Taurantha
Trd.	= Taurodium
Tdz.	= Telidezia
Tp.	= Telipogon
Tlt.	= Telisterella
Tem.	= Templeara
Thra.	= Teohara
Tbc.	= Tetrabaculum
Tbg.	= Tetrabroughtanthe
Ttct.	= Tetracattleya
Tcy.	= Tetracyclia
Ttdm.	= Tetradiacrium
Ttka.	= Tetrakeria
Tta.	= Tetralaelia
Ttps.	= Tetraliopsis
Ttr.	= Tetrallia
Ttma.	= Tetramicra
Ttt.	= Tetrarthron
Tyc.	= Tetrasychilis
Tttna.	= Tetratonia

(continued)

Trn.	= Tetronichilis
Th.	= Thaiara
Thy.	= Thayerara
The.	= Thecopus
Tcs.	= Thecostele
Ttp.	= Thecostelopus
Tlb.	= Thelybaculum
Tly.	= Thelychiton
Thel.	= Thelymitra
Typ.	= Thelypilis
Tyr.	= Thelyrillia
Thsra.	= Thesaera
Thmpa.	= Thompsonara
Tho.	= Thorntonara
Thr.	= Thouarsara
Tx.	= Thrixspermum
Thu.	= Thunia
Tnl.	= Thunilla
Thw.	= Thwaitesara
Tic.	= Ticoglossum
Tld.	= Toladenia
Tos.	= Tolassia
Tgz.	= Tolguezettia
Tnc.	= Toloncettia
Tln.	= Toluandra
Tun.	= Tolucentrum
Tgl.	= Toluglossum
Tolu.	= Tolumnia
Tmp.	= Tolumnopsis
Tut.	= Tolutonia
Tmz.	= Tomazanonia
Tod.	= Tomoderara
Trp.	= Trachypetalum
T.	= Traunsteinera
Trta.	= Trautara
Trcv.	= Trevorara
Try.	= Treyeranara
Trl.	= Triaristella
Trias	= Trias
Tspt.	= Triaspetalum
Tphm.	= Triasphyllum
Tri.	= Trichachnis
Tss.	= Trichassia
Tcn.	= Trichocenilus
Ths.	= Trichocensiella
Trt.	= Trichocentrum
Tc.	= Trichoceros
Tchch.	= Trichochilus
Tpy.	= Trichocidiphyllum
Trcdm.	= Trichocidium
Tyd.	= Trichocyrtocidium
Tdza.	= Trichodezia
Trgl.	= Trichoglottis
Tnn.	= Trichononcos
Trnps.	= Trichonopsis
Tric.	= Trichopasia
Trpla.	= Trichopilia
Tpgn.	= Trichopogon
Trcps.	= Trichopsis
Trla.	= Trichorella
Tsx.	= Trichosalpinx
Tcm.	= Trichosma
Tht.	= Trichostele
Trst.	= Trichostylis
Trcv.	= Trichovanda
Tvv.	= Trichovanvanda
Tyt.	= Tricyrtochilum
Tr.	= Tridactyle
Trgca.	= Trigolyca
Trgdm.	= Trigonidium
Trmn.	= Trilumna
Tcd.	= Triodoncidium
Tsla.	= Trisuloara
Tphnt.	= Trophianthus
Tpl.	= Tropilis
Trud.	= Trudelia
Tru.	= Trudelianda
Tsa.	= Tsaiara
Ts.	= Tsaiorchis
Tsi.	= Tsiorchis
Tst.	= Tsubotaara
Tbcm.	= Tubaecum
Tct.	= Tubecentron
Tbl.	= Tuberella
Tblm.	= Tuberolabium
Tpc.	= Tuberoparaptoceras
Tuck.	= Tuckerara
Tna.	= Tunstillara
Tbwa.	= Turnbowara
Twr.	= Twuara
Tylo.	= Tylochilus
Tyl.	= Tylostylis
Uml.	= Umlandara
Upta.	= Uptonara
Urc.	= Urochilus
Uro.	= Urostachya
Vach.	= Vacherotara
Val.	= Valinara
Vacl.	= Vanachnochilus
Vnsta.	= Vanalstyneara

(continued)

Vtp.	= Vananthopsis		Vrm.	= Veramayara
Vnc.	= Vanascochilus		Vbn.	= Verboonenara
Vcp.	= Vancampe		Vrml.	= Vermeulenia
Vct.	= Vanchoanthe		Ver.	= Vervaetara
V.	= Vanda		Vyr.	= Veyretia
Va.	= Vandachnanthe		Vnt.	= Vicentelara
Vchns.	= Vandachnis		Vnk.	= Vinckeara
Van.	= Vandachostylis		Vpda.	= Viraphandhuara
Vdcy.	= Vandacostylis		Vrs.	= Vitebrassonia
Vand.	= Vandaecum		Vtl.	= Vitechilum
Vdnps.	= Vandaenopsis		Vtc.	= Vitecidium
Vp.	= Vandaeopsis		Vit.	= Vitekorchis
Vths.	= Vandaeranthes		Vkt.	= Volkertara
Vps.	= Vandanopsis		Vbm.	= Vonbismarckara
Vt.	= Vandanthe		Vri.	= Vriesara
Vdla.	= Vandantherella		Vuylck.	= Vuylsteckeara
Vdr.	= Vandantherides		Vuyl.	= Vuylstekeara
Vc.	= Vandarachnis		Wai.	= Waibengara
Vth.	= Vandathera		Wlra.	= Wailaiara
Vns.	= Vandensonides		Wll.	= Wailaihowara
Vwga.	= Vandewegheara		Wre.	= Waireia
Vnd.	= Vandirea		Wrna.	= Waironara
Vf.	= Vandofinetia		Wyg.	= Waiyengara
Vfds.	= Vandofinides		Wlb.	= Wallbrunnara
Vg.	= Vandoglossum		Wrs.	= Warasara
Vgt.	= Vandoglottanthe		Wba.	= Warburtonara
Vdp.	= Vandopirea		Wtd.	= Warcatardia
Vdpsd.	= Vandopsides		Whb.	= Warchaubeanthes
Vdps.	= Vandopsis		Wcy.	= Warchlerhyncha
Vv.	= Vandopsisvanda		Wzt.	= Warczatoria
Vdts.	= Vandoritis		Wzb.	= Warczebardia
Vgm.	= Vanglossum		Wzr.	= Warczerhyncha
Vl.	= Vanilla		Wcz.	= Warczewiczella
Veg.	= Vanlaenoglottis		Wws.	= Warczewscaphe
Vnr.	= Vannerara		Wnra.	= Warneara
Vpl.	= Vanphalanthe		Wrn.	= Warnerara
Vqt.	= Vanquetia		Wra.	= Warrea
Vst.	= Vanschista		Wtr.	= Warscatoranthes
Vsp.	= Vanstauropsis		W.	= Warscewiczella
Ves.	= Vansteenisara		Wwz.	= Warszewiczara
Vtn.	= Vantonia		Wtb.	= Watanabeara
Vrp.	= Vantrichopsis		Wts.	= Watsonara
Vvd.	= Vanvanda		Wbb.	= Weberbauerara
Vapd.	= Vappaculum		Wel.	= Wellesleyara
Vap.	= Vappodes		Wgg.	= Wengangara
Vrg.	= Vargasara		Wsta.	= Westara
Vasco.	= Vascostylis		Whi.	= Whitinara
Vnra.	= Vaughnara		Wmr.	= Whitmoorara
Vja.	= Vejvarutara		Wig.	= Wiganara
Vrr.	= Veraara		Wbg.	= Wilburachangara

Wbchg.	= Wilburchangara		Zdp.	= Zelencidopsis
Whm.	= Wilhelmara		Znt.	= Zelenettia
Wil.	= Wiliaara		Zgs.	= Zelengomestele
Wlk.	= Wilkara		Zel.	= Zelenkoa
Wknsra.	= Wilkinsara		Zka.	= Zelenkoara
Wlm.	= Williamara		Zed.	= Zelenkocidium
Wlc.	= Williamcookara		Zgd.	= Zelglossoda
Wmp.	= Williampriceara		Zht.	= Zellahuntanthes
Wls.	= Williamsara		Zmg.	= Zelomguezia
Whw.	= Williehowara		Zds.	= Zeloncidesa
Wmt.	= Wilmotteara		Zts.	= Zeltonossum
Wils.	= Wilsonara		Zgz.	= Zelumguezia
Wgfa.	= Wingfieldara		Zyd.	= Zelyrtodium
Win.	= Winnara		Zo.	= Zootrophion
Wse.	= Wiseara		Zbd.	= Zygobardia
With.	= Withnerara		Zbt.	= Zygobatemania
Woj.	= Wojcechowskiara		Zbm.	= Zygobatemannia
Wly.	= Wolleydodara		Zyo.	= Zygocalyx
Woo.	= Wooara		Zcst.	= Zygocaste
Wdwa.	= Woodwardara		Zcl.	= Zygocastuloa
Wr.	= Wrefordara		Zy.	= Zygocidium
Wrg.	= Wrigleyara		Zcx.	= Zygocolax
Wud.	= Wudhikanakornara		Zyg.	= Zygodendrum
Wut.	= Wuttiphanara		Zdsnth.	= Zygodisanthus
Xar.	= Xaritonia		Zgm.	= Zygogardmannia
Xra.	= Xerriara		Zgx.	= Zygolax
Xip.	= Xiphosium		Zglm.	= Zygolum
Xyl.	= Xylobium		Zmt.	= Zygomatophyllum
Yhra.	= Yahiroara		Zma.	= Zygomena
Yam.	= Yamadaara		Zmz.	= Zygomenzella
Yamad.	= Yamadara		Zgc.	= Zygoncidesa
Yap.	= Yapara		Zga.	= Zygoneria
Yra.	= Yeeara		Zsr.	= Zygonisatoria
Ypga.	= Yeepengara		Zns.	= Zygonisia
Ybk.	= Yeohbokara		Zba.	= Zygopabstia
Ymn.	= Yinmunara		Z.	= Zygopetalum
Yin.	= Yinwaiara		Zpn.	= Zygophinia
Yit.	= Yithoeara		Zcha.	= Zygorhyncha
Ynra.	= Yoneoara		Zcp.	= Zygoscaphe
Yzwr.	= Yonezawaara		Zspm.	= Zygosepalum
Ygt.	= Youngyouthara		Zsp.	= Zygosepella
Ypt.	= Ypsilactyle		Zsc.	= Zygosepescalum
Yp.	= Ypsilopus		Zst.	= Zygosteria
Ysfra.	= Yusofara		Zsts.	= Zygostylis
Ywg.	= Yuwengangara		Zgt.	= Zygotoria
Zwn.	= Zelawillumnia		Zwr.	= Zygowarrea
Zlm.	= Zelemnia		Zzl.	= Zygozella
Zlc.	= Zelenchilum			
Zcs.	= Zelenchostele			
Zct.	= Zelencidiostele			

Glossary

Aerial root   Any root emanating from above the growing medium. Many epiphytic orchids produce aerial roots due to their growing openly on trees.

Alba   The white form of a flower.

Alliance   A group of related genera in which crossbreeding between the genera is possible.

Anther cap (anther)   In orchids, a cap of tissue that houses the pollen masses at the top of the column.

Asymbiotic culture   A technique developed by Lewis Knudson for the germination of orchid seeds in a nutrient-rich medium without the aid of a mycorrhizal fungus.

Autogamous   Self-fertilizing.

Backbulb   In a sympodial orchid, an old pseudobulb that has finished its growing cycle. It can be used to propagate a new plant from one of its dormant buds after removal from the plant.

Bifoliate   Having two leaves from a single pseudobulb.

Bract   A small modified leaf that generally supports a flower.

Bud   Term for a flower before it opens and can also refer to a new growth or leaf.

Calceolate   Slipper-like form, such as the shape of a *Paphiopedilum*.

Callus (pl. Calli)   A ridge-like, waxy growth on the labellum of an orchid.

CAM   Crassulacean Acid Metabolism. A form of photosynthesis in which the leaf stomata of a plant close during the day to avoid water loss and open at night in order to acquire carbon dioxide, which is stored overnight as malic acid. During daylight, this latter is converted back to carbon dioxide and conventional photosynthesis takes place. Many orchid species employ this type of photosynthesis in regions where water is scarce.

Cane   A very elongated pseudobulb usually having nodes, as in the case of certain *Dendrobiums*.

Capsule   The closed seedpod of an orchid plant, sometimes containing thousands or even millions of seeds.

Caudicles   Thin strands that attach pollinia to each other or to the viscidium. Part of the pollen mass and produced within the anther. Not to be confused with the stipe.

CITES   Convention on International Trade in Endangered Species. The multinational agreement that lists which plant and animal species are considered endangered and the rules by which their trade is governed.

Clinandrium   A small hollow at the top of the orchid column where the pollinia reside.

Clone   The offspring of a plant derived by cutting or tissue culture, thereby producing an identical copy of its parent.

Column   A single columnar structure containing both the male and female sexual parts of an orchid. This is one distinguishing feature of orchids from other flowering plants. The technical name for the column is gynostemium.

Cotyledon   The leaf or leaves of a seed embryo. Orchids have a single cotyledon.

Cross   The offspring that result from the transfer of pollen from one plant to the flower of

J.L. Schiff, *Rare and Exotic Orchids*, https://doi.org/10.1007/978-3-319-70034-2

another, although the flowers of a plant can also be crossed with those of itself; also refers to the act itself.

Crown The central part of the leaf coil in a monopodial orchid such as *Phalaenopsis*, from which new growth is produced.

Cultivar In orchids, a particularly outstanding example of a species or hybrid that could merit an award, designated by single quotes around its name.

Damping off When seedlings die off due to various pathogens.

Deciduous The seasonal shedding of leaves after growth and maturity; not evergreen.

Division Vegetative propagation of a new plant from an existing one made by cutting the rhizome of a sympodial orchid into two or more parts possessing pseudobulbs.

Diploid Having the standard number of two sets of chromosomes; also known as 2 N. Humans are diploid and so are most orchids, although not all of the latter. Some can be triploid (3 N) or tetraploid (4 N) in special circumstances.

Dorsal sepal The uppermost sepal of an orchid flower.

Embryo At the mature stage in orchids, it represents the cellular embodiment of the orchid plant within the testa before it has germinated. However, it lacks the endosperm required for germination.

Endemic Found in a restricted region, country or island.

Endosperm Tissue containing vital nutrients of starches and proteins surrounding a plant seed's embryo, used for its initial phase of development. This tissue is lacking in orchid seeds, hence their need for a mycorrhizal fungus to germinate.

Epichile The lower portion of the lip of certain orchids such as *Coryanthes* and *Stanhopea*.

Epiphyte A plant that grows on another plant for support but takes no nourishment from its host and hence is not parasitic; adj. Epiphytic. Nutrients are obtained from the air, rain, and organic debris.

Eye The bud of a growth that can develop into new a growth.

Flask A glass container used for the laboratory germination of orchid seeds and seedlings. The process is called flasking.

Floriferous A plant that is freely flowering.

Foot candle A measure of light intensity for the growing of plants; it is the illumination produced by a candle at a distance of one foot on a 1 square foot surface. 1 footcandle = 1 lumen/ft^2.

Foliar spray A mixture of plant nutrients in water to spray on leaves and be absorbed by the stomata. For CAM orchids, this is best done in the evening when the stomata are open.

Genus (pl. Genera) A grouping of related species that possess similar characteristics and a presumed common ancestry.

Grex The second name used to describe a particular hybrid cross, such as *Cattleya* Michelle Obama.

Gynostemium The technical term for the column of an orchid (see column).

Habitat The natural wild environment in which a plant grows.

Haploid Having a single set of chromosomes, denoted by N. Compare with diploids, which are 2 N. Orchid haploids have been grown artificially.

Hybrid The offspring resulting from a cross between two different species (a primary hybrid) or two different hybrids (complex hybrid).

Hyphae Thin filaments of various fungi mycelium (the mycelium being the vegetative part of the fungus, not the fruiting body that is usually seen) that provide the orchid seed embryo with nutrients necessary for the seed's germination.

Hypochile A bulbous part of the labellum in such orchids as *Coryanthes* and *Stanhopea*.

Inflorescence The flowering portion of the orchid that can take various different forms (panicle, raceme, scape, etc.); commonly referred to as a "spike" regardless of form.

Indigenous Native to a particular region or country as opposed to introduced.

Intergeneric Between two or more genera generally in the context of hybridization.

In vitro Taking place in an environment outside an organism, as in a test tube or petri dish.

Keiki A Hawaiian word for "baby" or "child" that refers to a plantlet produced (asexually) by an orchid plant, common to cane-like *Dendrobiums* and also found on *Epidendrums*, *Phalaenopsis*, *Vandas*, and *Catasetum*. When roots are sufficiently developed, they can be detached and grown as separate plants.

Labellum A modified petal on an orchid flower that generally plays a role in the pollination process. Notable for its distinguished appearance, which can also serve as a landing platform for pollinators.

Lateral sepals The two lowermost sepals extending to the sides, beneath the two petals as opposed to the uppermost dorsal sepal.

Lip Common term for the orchid labellum.

Lithophytic Used to describe the growth habits of plants that grow affixed to rocks. Such plants are lithophytes. Their nourishment is derived from the air, rain, mosses, and decayed debris.

Medium The potting mixture being used in the orchid pot. For non-terrestrial orchids, it can be organic such as fir or pine bark, coconut husks, sphagnum moss, tree fern fibers, or inorganic such as perlite, pumice, as well as various combinations of these materials.

Mericlone A plant propagated artificially in a laboratory by meristem tissue culture. The resulting plant is an exact copy (clone) of the original.

Meristem An active growing area of undifferentiated cells of the plant from which tissue is taken in order to produce a mericlone.

Mesochile The middle portion of the labellum between the hypochile (top) and epichile (bottom) of a complex labellum as found on *Coryanthes* and *Stanhopea* orchids.

Micropropagation The creation of new orchids in the laboratory by such methods as meristem tissue propagation and seed culture.

Monopodial The form of orchid vegetative growth along a single stem that grows continually upward with the leaves forming alternately on each side of the stem. Much less common than sympodial orchids with examples being, *Sarcochilus*, *Vanda*, and *Vanilla*, orchids.

Monocotyledons Plants whose seeds typically contain a single embryonic leaf (cotyledon), as is the case with orchids. Referred to as monocots.

Myco-heterotrophy A special symbiotic relationship between certain orchids and fungi in which the orchid plant obtains all or part of its nutrients from the fungi rather than from photosynthesis, typically due to lack of leaves.

Mycorrhizal fungus The type of fungus that lives in a symbiotic relationship with an orchid plant. Suitable mycorrhizal fungi are necessary for orchid seed germination in nature.

Natural hybrid A hybrid that occurs in nature and not by artificial means.

Nectary A tubular spur at the back of an orchid flower that secretes and stores nectar.

Node A joint or notch on a flower stem or pseudobulb from which leaves, flowers, roots, or an inflorescence can originate. The latter is often the case when a *Phalaenopsis* stem is cut just above a node in order to encourage another flower stem to propagate from the same node.

Nonresupinate The word itself means "upside down," and in the case of orchids, it means those plants whose flower lips are positioned uppermost; most orchid flowers are resupinate with the lip at the bottom.

Orchids Members of the Orchidaceae family of monocotyledons, a major group of flowering plants. With approximately 28,000 species, they have a greater diversity and more specialized means of pollination than any other flowering plants.

Osmophores Special glands of a flower responsible for producing its scent.

Ovary The part of the orchid flower that produces the seed after the flower has been pollinated.

Panicle A many-branched inflorescence.

Pedicel The small stalk bearing an individual flower within an inflorescence.

Peloric A genetic mutation in orchids in which the two lateral petals or sepals take on some of the characteristic appearance of the lip.

Petal One of the three components of the orchid flower that are positioned alternately between the three sepals; one distinguished petal is modified into the labellum.

Pheromone A chemical compound given off by an insect (or animal) in order to affect the behavior of other members of its species, such as attracting a mate of the opposite sex.

Photosynthesis The plant process by which chlorophyll in the leaves transforms sunlight into chemical energy that is used to synthesize sugars from carbon dioxide (CO_2) absorbed from the air through leaf stomata, and from water taken up by the roots. Oxygen (O_2) is vented out as a waste by-product through the stomata.

Pod Another term for the seedpod or capsule.

Pollinarium (pl. pollinaria) The whole structure consisting of pollinia, stipe, and viscidium, forming the pollination unit.

Pollination When the pollen of one orchid flower is placed on the stigma of another, whether by natural or artificial means.

Pollinium (pl. pollinia) In orchids, a coherent packet of pollen grains found under the anther cap at the top of the column, unlike other flowers where the pollen grains are a fine powder. In general, orchid flowers have either 2, 4, 6, or 8 pollinia. Pollinia come in two basic types, soft and mealy or hard and waxy.

Polyploid A plant with more than the normal two sets of chromosomes (2 N) such as triploid (3 N) and tetraploid (4 N), etc.

Protocorm Small tuber-shaped bodies that result from the germination of orchid seeds that is the embryonic form of the orchid plant prior to leaves and roots being formed.

Pseudobulb Thickened portion of the stem that branches off the rhizome of sympodial epiphytic orchids. Stores water, minerals, carbohydrates and other nutrients for the plant's vegetative development, including reproduction. It is not a bulb in the true sense of the word (like in a tulip bulb).

Pseudocopulation The actions of a male insect attempting to copulate with an orchid flower it has been deceived into believing is a female of its own species. Leads to the possible pollination of the flower.

Raceme An inflorescence that is unbranched, bearing flowers along its length. Most orchids have a racemose inflorescence.

Resupinate In orchids, those flowers whose lips are positioned at the bottom of the flower as is the case with the vast majority of orchids. Non-resupinate flowers have their lips positioned uppermost and appear upside-down.

Rhizome A woody, horizontally growing stem of sympodial orchids that produces roots and new vertical plant growth. Rhizomes will often branch, allowing for the multiplication of orchid plants by division.

Rostellum An outgrowth that separates the pollinia from the stigma and inhibits self-pollination.

Scape An unbranched inflorescence topped with a single flower, as is the case for many *Paphiopediums*.

Seed The mature embryo enclosed by the testa. Most orchid seeds are minute in size, and there can be thousands to millions of seeds in a single (seed)pod.

Selfing Making a cross with a single flower or two different flowers on the same plant.

Semialba A white flower that has a colored lip.

Sepal One of the three components of the orchid flower that are positioned alternately between the three petals, typically starting with the uppermost dorsal sepal (resupinate flowers).

Sheath A modified leaf structure surrounding some parts of the orchid plant such as a flower spike or pseudobulb.

Sib cross, Sibling cross The cross-pollination of two plant siblings.

Sibling An orchid that is related to another orchid, having come from the same seedpod.

Species A group of living things that share common characteristics distinguishing them from other such groupings. Forms the standard classification unit of orchid genera.

Specimen plant One that has been grown to a large size rather than being divided.

Spike General term used to refer to any unbranched inflorescence; technically an unbranched inflorescence stem having flowers without pedicels (stalkless).

Spur A hollow tubular extension from the base of the lip usually bearing nectar for potential pollinators.

Staminode   A small shield-like appendage behind which lie sexual parts of a *Paphiopedilum* or *Cypripedium*.

Stigma   The female part of the flower, generally a shallow depression, that is receptive to the pollen and located on the column.

Stipe   A stalk-like structure connecting pollinia to the viscidium, variable in length and thickness and composed of tissue derived from the column. Compare with caudicles.

Stomata (sing. Stoma)   Minute apertures on a leaf surface that open and close to allow the passage of carbon dioxide in and oxygen and water out during photosynthesis.

Symbiotic culture   The germination of orchid seeds through the application of a mycorrhizal fungus.

Sympodial   One of the two types of orchid growth in which new growth arises vertically along a laterally spreading rhizome.

Synsepal   The fused lateral sepals of some orchids such as *Restrepia* and *Paphiopedilum*.

Tepal   A term denoting both the sepals and petals but excluding the lip.

Terete   Leaves that are elongated and pencil-shaped as in some species of *Dendrobium* (*Dockrilla*).

Terrestrial   In orchids, growing in the ground or in the debris on a forest floor.

Testa   Tough protective seed covering of an embryo.

Throat   The innermost portion of a tubular orchid lip.

Tetraploid   Term for a plant having four sets of chromosomes (4 N) due to a genetic aberration in contrast to a normal diploid plant with two sets of chromosomes (2 N). The plants are very important in modern breeding due to the double influence of their genes.

Tissue culture   Artificial propagation of plants in a laboratory by means of meristem culture.

Triploid   Having three sets of chromosomes (3 N), generally the result of crossing a diploid (2 N) with a tetraploid (4 N). Most modern *Cymbidium* hybrids are triploids.

Tuber   Thickened subterranean masses used by terrestrial orchids to store nutrients.

Type specimen   The particular species that typifies a given genus.

Unifoliate   Having a single leaf per pseudobulb.

Variety   Plant having minor variations from the type species.

Velamen   Protective coating on orchid roots often a gray-white in color, which absorbs water and nutrients for the plant and protects the root cortex from drying out.

Viscidium   A viscid disk connected to the pollinia that attaches to a pollinator in some orchids.

Zygomorphic   Bilateral symmetry about a vertical axis whereby the left side is the mirror image of the right side. A characteristic exhibited by all orchid flowers.

Bibliography

Books

John Alcock, *An Enthusiasm for Orchids – Sex and Deception in Plant Evolution*, Oxford University Press, 2006.

Joseph Arditti, *Fundamentals of Orchid Biology*, Wiley, 1992.

Luigi Berliocchi, *The Orchid: in Lore and Legend*, Timber Press, 2000.

Charles Darwin, *The Various Contrivances by Which Orchids are Fertilized by Insects and On The Good Effects of Intercrossing*, John Murray, London, 1st ed., 1862, 2nd ed., 1877.

Eric Hansen, *Orchid Fever – A Horticultural Tale of Love, Lust, and Lunacy*, Vintage Books, 2001.

Retha Edens-Meier, Peter Bernhardt eds., *Darwin's Orchids: Then and Now*, University of Chicago Press, 2014.

Jim Endersby, *Orchid – A Cultural History*, University of Chicago Press, 2016.

Roman Kaiser, *The Scent of Orchids*, Elsevier Science, 1992.

Isobyl la Croix, *The New Encyclopedia of Orchids*, Timber Press, 2008.

Susan Orlean, *The Orchid Thief*, Vintage Books, 2000.

Craig Pittman, *The Scent of Scandal: Greed, Betrayal, and the World's Most Beautiful Orchid*, University of Florida Press, 2012.

Merle A. Reinikka, *A History of the Orchid*, Timber Press, 1995.

Arthur Swinson, *Frederick Sander: The Orchid King*, Hodder and Stoughton, 1970.

Eng Soon Teoh, *Medicinal Orchids of Asia*, Springer, 2016.

N.A. Van Der Cingel, *An Atlas of Orchid Pollination: European Orchids*, CRC Press, 2001.

Websites

BlueNanta. http://bluenanta.com/

Hortus Orchis. http://www.hortusorchis.org/en/

Internet Orchid Species Photo Encyclopedia. http://www.orchidspecies.com

Kew Gardens World Checklist of Selected Plant Families. http://apps.kew.org/wcsp/prepareChecklist.do?checklist=selected_families%40%40114240420170734018/

The Plant List. http://www.theplantlist.org/1.1/browse/A/Orchidaceae/

Tropicos (Missouri Botanical Gardens). http://www.tropicos.org

Index

Proper orchid abbreviations have not been used for purposes of the Index.

MIX
Papier aus verantwortungsvollen Quellen
Paper from responsible sources
FSC® C105338

If you have any concerns about our products,
you can contact us on
ProductSafety@springernature.com

In case Publisher is established outside the EU,
the EU authorized representative is:
Springer Nature Customer Service Center GmbH
Europaplatz 3, 69115 Heidelberg, Germany

Printed by Libri Plureos GmbH
in Hamburg, Germany